T0338995

DISTILLED

DISTILLED

A Natural History of Spirits

ROB DESALLE &
IAN TATTERSALL

Illustrated by Patricia J. Wynne

With contributions from
Miguel A. Acevedo, Sergio Almécija, Angélica Cibrián-Jaramillo,
Tim Duckett, Joshua D. Englehardt, Michele Fino, Michele Fontefrancesco,
América Minerva Delgado Lemus, Pascaline Lepeltier, Christian McKiernan,
Mark Norell, Susan Perkins, Bernd Schierwater, Ignacio Torres-García,
Alex de Voogt, and David Yeates

Yale UNIVERSITY PRESS/NEW HAVEN & LONDON

Published with assistance from the Louis Stern Memorial Fund.

Frontispiece photo: (altered) iStock.com/gresei.

Designed by Mary Valencia.
Set in Adobe Text Pro type by Integrated Publishing Solutions.
Printed in the United States of America.

Library of Congress Control Number: 2021946782
ISBN 978-0-300-25515-7 (hardcover : alk. paper)

A catalogue record for this book is available from the British Library.

This paper meets the requirements of ANSI/NISO Z39.48-1992 (Permanence of Paper).

10 9 8 7 6 5 4 3 2 1

To Patrick McGovern,
with appreciation from all who share his passion for
the ancient history of alcoholic beverages

Contents

Preface

After our two books *A Natural History of Wine* and *A Natural History of Beer,* a companion volume on distilled spirits might seem almost inevitable. But it is no afterthought. It is true that fermented beverages like wine and beer were around for millennia before some creative soul figured out how to make them into highly alcoholic spirits. But when distillation came about, this formidable technological development didn't just pump up the alcohol; it radically broadened the range of gustatory experiences available to drinkers. That is because ethanol is far from being a tasteless substance, as is often believed. Instead, once it rises above a concentration of about 20 percent by volume, well above anything you would expect to find in a conventional beer or wine, it imparts a subtly bittersweet and uniquely mouth-filling quality that has yet to be replicated by any other substance. Often described accurately, if sometimes a little unfairly, as a "burning" sensation, this sensory quality has been ingeniously exploited by distillers to provide an almost infinite variety of taste experiences, the breadth of which far surpasses those offered by drinks made simply from fermented grape or grain.

Just as in our earlier books, we will look at our subject, distilled spirits, through the lens of natural history, positioning it within the diverse contexts of evolution, ecology, history, primatology, molecular biology, physiology, neurobiology, chemistry, and even astrophysics—all in the hope of enhancing your appreciation of a liquid that hopefully reposes in a glass nearby. But we also depart a bit from our earlier format. Spirits are in certain respects trickier for the analyst than beers or wines because, particularly with barrel aging, spirits derived from vastly different sources can converge significantly over time, just as spirits within the same general category can develop vastly different characteristics. Because of the sheer variety of distilled spirits on the market, then, both in terms of major categories and of regional variations on the distilling theme, we have called on several knowledgeable colleagues to contribute or collaborate with us on some chapters that deal with specific spirits or drink types. Our deepest thanks go to all of these colleagues, whose expertise and enthusiasm have made this volume far more entertaining and authoritative than it otherwise would have been.

As this final volume of our "natural history of alcohol" series appears, it is also appropriate to acknowledge the many contributions of Patrick McGovern, genial colleague and preeminent authority on early alcoholic drinks. Everyone who writes or cares about the ancient history of alcoholic beverages owes a huge debt to "Doctor Pat," and we owe even more, because all three volumes of our series were vastly improved by his careful readings and commentaries. It is thus a pleasure as well as an honor to have dedicated this one to him. Any lingering deficiencies are, of course, entirely our own.

Our thanks also go, as ever, to our wonderful editor Jean Thomson Black, who has enthusiastically supported this series since its very beginning. Without her active engagement we could never have carried it through, and the process certainly wouldn't have been as much fun. The same applies to our illustrator, Patricia Wynne, without whom it would by now be almost unthinkable for us to contemplate taking on a project like this. At the Yale University Press, we are also deeply indebted to Margaret Otzel, Elizabeth Sylvia, and Amanda Gerstenfeld for keeping the project on track with the greatest of good humor, to Julie Carlson for smoothing everything out with her careful copyedit, and to Mary Valencia for her elegant design.

The world of spirits is, of course, so broad and deep that we would hardly have dared to take on the subject without the encouragement and support of many other fellow spirits lovers and, in some cases, mentors. Besides those who have joined us as authors (with a special extra shout-out to Alex de Voogt), those aficionados include Will Fitch, Robin Gilmour, Marty Gomberg, Jeanne Kelly, Jakob Köllhofer, Chris Kroes, Brian Levine, Mauri Rosenthal, Nancy Taubenslag, and Sarah Weaver. Thank you all. And, as ever, our gratitude and affection go to our wives, Erin and Jeanne, not least for their forbearance and good humor during the production of this entire trilogy.

Spirits in History and Society

1

Why We Drink Spirits

If a monkey could enjoy the products of fermentation, we thought, why not us? So we were a little disappointed to learn that despite the glass monkeys decorating the bottle in front of us, its contents actually had no connection to our primate relatives. "Monkey shoulder" is, apparently, a condition that afflicts Scottish maltsters when they have shoveled their grain too energetically. Nonetheless, we were eager to sample those contents, and our optimism was rewarded. The malt shovelers who had made this bottle possible had helped to create a satisfyingly smooth, blended Scotch whisky. A light straw color, it offered faintly peaty aromas, enhanced with a whiff of dried apricots and a whisper of fresh-mown grass. On the palate it was seductively smooth, with a heavy accent of salted caramel and a hint of vanilla. This was a classic blend, with a lingering finish and a harmonious integration of flavors. We hoped that no humans had been harmed in its production.

Everyone knows why they like distilled spirits, and why they drink them. If, indeed, they do like and drink them—because spirits, as the strongest alcoholic beverages on the market, with the most aggressive flavor profiles, take rather extreme positions on the sensory and alcoholic scales. Yet those who do appreciate the sensory complexities of spirits, and the sheer gustatory variety they offer us, are almost invariably enthusiasts. For them (and us), existence without this "water of life" would be frustratingly incomplete.

As sensory beings we are middling creatures, with neither the sharpest eyesight nor the keenest noses found in nature. But our sensory capacities are nonetheless astonishing, and they happen to be servants of a very unusual cognitive system: one that allows us to analyze our sensory inputs in a unique—and uniquely satisfying—way. We human beings crave any coherent synthesis we can achieve of our sensory and cerebral experiences, and at a very fundamental level this dynamic integration of senses and intellect is what makes the experience of drinking spirits so rewarding.

Still, there is no point denying that spirits also offer us the fastest way to get drunk. And although humans are clearly not the only creatures that get a buzz out of alcohol—everyone is surely familiar with those hilarious clips on YouTube that show elephants allegedly snookered on fermenting amarula fruits—humans are the only ones, as far as we know, who perceive the intrinsic uncertainties of life and are condemned to live with an explicit awareness of their own mortality. This knowledge places an existential burden on humankind that no other species has to bear. Members of our species are uniquely capable of worrying not only about what is happening to them right now, but also about things that will or might happen to them in the future. Since we know that our lives are uncertain and fraught with hazard, we tend to welcome anything that will help distance us from this unpleasant reality. Alcohol, through its relaxing and distracting effects, helps us keep that distance, and spirits offer both an effective and a sociable way of delivering that happy succor. What's more, on those hopefully more frequent occasions when we are not burdened with angst, we warm to the general lessening of inhibitions and the sense of connection with others that alcohol can provide. As long as excess is avoided, the benevolent effects of this remarkable molecule reliably prevail.

Patrick McGovern, the leading authority on ancient alcoholic beverages, has also pointed out some possible roles of those beverages in prehistoric times. He notes that, at a time when "synthetic drugs were not available and your life span, if you survived birth and childhood, was 20–30 years, a fermented beverage's health benefits were obvious—alcohol relieved pain, stopped infection, and seemingly cured disease." He points out that alcohol has always been easily administered by drinking, or by direct application to the skin, and he suggests that, in times before modern sanitation, the drinkers of fermented beverages (in preference to frequently tainted water) "lived longer and consequently reproduced more." Additionally, "herbal or tree resin compounds with medicinal properties could also be more easily dissolved in an alcoholic medium than in water," while "besides the sheer psychoactive delight in th[e] new-found beverages, fermentation produced more nutritious, sensorily appealing, and more preservable foods than the starting materials." What's more, in its role as a "social lubricant," alcohol served and still serves to "bring people together as a group by breaking down inhibitions between individuals." McGovern also stresses that the mind-altering effects of fermented beverages, not to mention the essential mysteriousness of fermentation itself, help explain how readily those beverages were "incorporated into religions around the world." Without a doubt, then, life in a world without alcohol would be very different from the one we know. And, to most of us at least, much less appealing.

All these positive aspects notwithstanding, *Homo sapiens* is perhaps most convincingly defined as a species compelled to take every excellent idea to a crazy extreme. Of all the alcoholic beverages out there, spirits are unquestionably the most readily abused. Society needs to address this serious issue rationally, even if it is clear that, in moderation, spirits effectively offer solace when needed and, under happier circumstances, unique sensory delights.

Another way to answer the question of *why* we drink spirits is to point out that we *can*. We tend to take our (limited) tolerance for alcohol pretty much for granted, but the plain fact is that ethanol, the alcohol in drinkable spirits, is toxic to most organisms. Indeed, some scientists reckon that the ancient ancestors of yeast, the microorganisms that produce the alcohol that spirits-makers distill today, initially made the stuff to poison the other tiny creatures with which they competed for ecological space. Ironically, even the hardiest of yeasts willingly poison themselves when the environmental level of alcohol created by winemakers and brewers exceeds about 15 percent by volume. And if any other organism wants to ingest alcohol and live, it needs to find a way to detoxify it.

In creatures like us, this detoxification is carried out by a specific class of enzymes known as alcohol dehydrogenases (ADHs). In humans, these molecules are called ADH1A, ADH1B (or ADH2), ADH1C (or ADH3), ADH4, ADH5, ADH6, and ADH7. The ADH1 molecules are found in all animals, and in humans they make up about 10 percent of the total enzymes of the liver. ADH4 molecules are present in the tissues of the tongue, as well as in the esophagus and stomach, and are thus the first enzymes to interact directly with the alcohol you consume. Alcohol dehydrogenases are responsible not only for breaking down alcohol, but also for maintaining in the cell a steady stream of a small molecule called NAD (nicotinamide adenine dinucleotide) that is used in energy production.

In 2015 scientists compared the distribution of ethanol-targeting ADH4s among a dozen species of primates, members of the diverse mammal order to which we humans also belong. Representing a broad cross-section of the primates, those species ranged from some rather distant cousins, the bushbabies of Africa and the aye-ayes of Madagascar, to monkeys from both the New and Old Worlds, to our close chimpanzee relatives and our own *Homo sapiens.* When they analyzed their results, the researchers discovered that some ten million years ago there was a dramatic switch in the human lineage, from an ancestral "ethanol inactive" form of ADH4 (the same one found in almost all other primates) to an "ethanol active" one. This change, which resulted from a single gene mutation, led to an amazing fortyfold increase in the body's ability to metabolize ethanol, and thus to neutralize it. Interestingly, this crucial change occurred *before* the separation of the lineages that led on the one hand to the modern African apes (the gorillas and chimpanzees), and on the other to us humans. This means that we share the powerful new enzyme with those apes (Figure 1.1).

So, what happened? Well, in evolution things don't have to happen for a reason. Mutations of the genes that compose our genetic code are happening all the time. These mutations are spontaneous, occurring merely as copying errors in the genetic code when cells divide. They consequently happen entirely at ran-

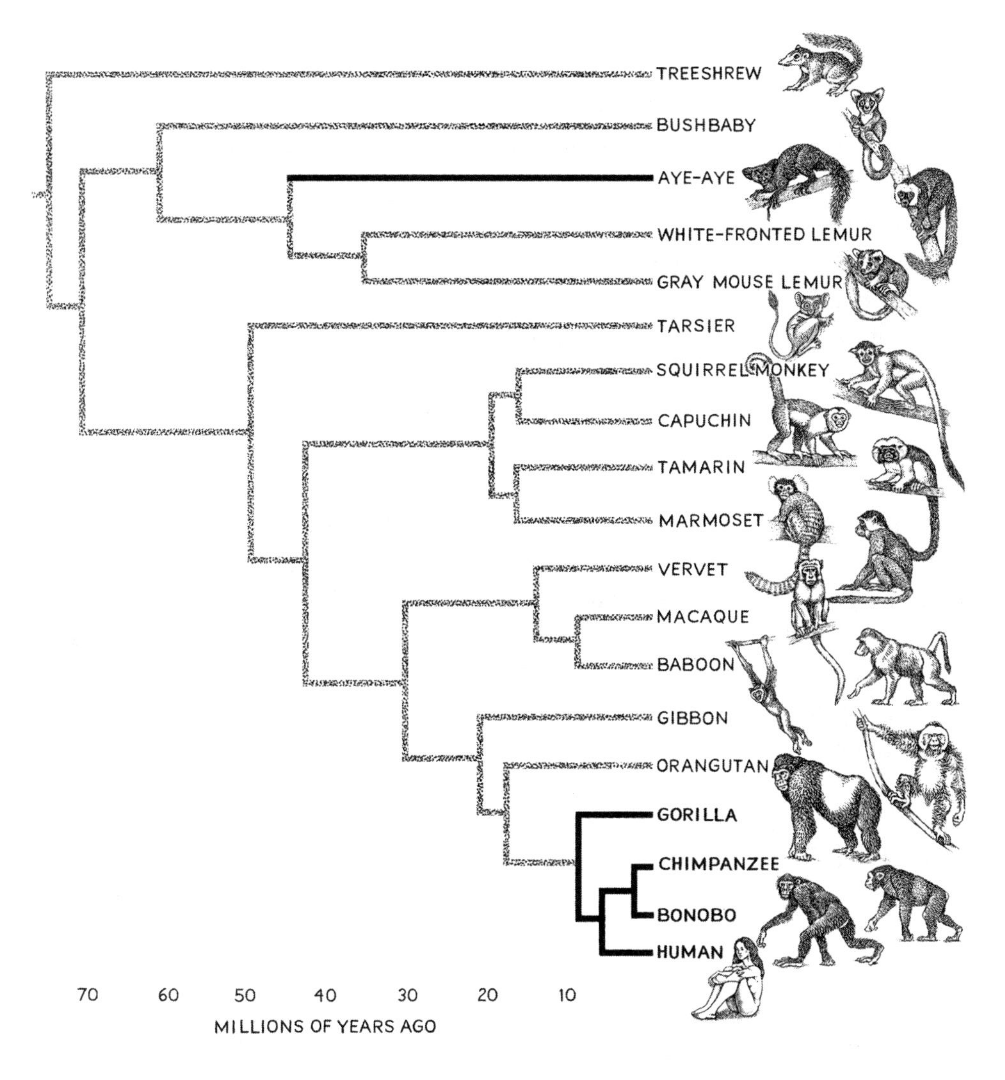

Figure 1.1. Tree showing the relationships among the primates tested for ADH4. Heavy lines indicate lineages with the ethanol-active form. Data from Carrigan et al. (2015).

dom, unrelated to any effects they might have, beneficial or otherwise. And whereas most of those changes will be eliminated quickly by natural selection, changed genes that don't get in the way too much may simply persist or, due to sheer chance, may actually spread in a population, especially a small one. In other words, genes don't have to be actively advantageous to survive, or even to become the norm. So it's perfectly likely that the new form of ADH4 arose merely by chance in the lineage that later split to give rise to the African apes and humans, and, because it posed no disadvantages—and it's very hard to imagine how it would have—subsequently spread by random processes in the small parental population.

It is also possible that the new enzyme might actually have been advantageous to its possessors, and that it became fixed in the ancestral population for that reason. Indeed, in the introductory chapter to their book *Alcohol and Humans: A Long and Social Affair,* the anthropologists Kimberley J. Hockings and Robin Dunbar propose that the new ability to deal with naturally occurring alcohol might actually have saved our ancient ancestor from extinction. They argue that around ten million years ago the apes (members of a formerly very successful and diverse forest-living group that was by then in decline) were coming under increasing pressure from the flourishing cercopithecine monkeys. Specifically, those apes found themselves in competition with those monkeys for the ripe fruit that composed the core of their diet. According to Hockings and Dunbar, the monkeys held the digestive advantage in this competition because they were able to process fruits at all stages of ripeness. But there was one source of fruit for which they could not compete, namely the overripe stuff that had fallen to the forest floor and that lay there gently fermenting, as rotting fruit will do once invaded by wild yeasts. That fermentation produced alcohol—a new ingredient that, the argument goes, placed this particular resource off-limits to the monkeys, which lacked the magic enzyme needed to deal with it in any quantity. The newly alcohol-tolerant human-chimpanzee ancestors could therefore munch on those fermenting fruits as much as they liked.

Well, that's a very nice story, and one would like it to be true. But since fermenting fallen fruit could only ever make up a small proportion of the diet of even the most dedicated large-bodied frugivore, and because chimpanzees today (let alone humans) have pretty varied diets, we feel it's a bit dubious to propose a causal link between the new enzyme and our ancestral diet. It is also clear that the remote human ancestor in question lived a very long time before our own precursors had become the generalist omnivores that their descendants remain, which also suggests that the new ethanol-active ADH enzyme was not associated with anything that humans or their precursors did, either recently or long ago. That is, while the new and improved alcohol dehydrogenase might well have helped one lineage of ancient apes to get through some tough times, ADH is unlikely to have originated in that context.

Still, whatever the actual circumstances of this notable physiological innovation may have been, we can confidently conclude that, down the line, its ultimate effect was to "preadapt" more recent humans to handle ethyl alcohol. And millions of years later, when humans figured out fermentation technology, they were positioned to take full advantage of it.

Fermentation requires sugar, and sugars abound in nature, largely as a result of plants' efforts to attract mobile creatures to eat their fruit, and thus to disperse their seeds. Nobody knows for sure when humans began to deliberately ferment sugars. That's because prior to the invention of pottery early in the Neolithic (New Stone Age) period, around nine thousand years ago, any fermentation vessels would necessarily have been made from perishable materials, of which we obviously now have no evidence. There's one possible exception: the archaeologist Oliver Dietrich and his colleagues once suggested that some hewn limestone basins at Pre-Pottery Neolithic levels of the Turkish site of Göbekli Tepe might have been used for brewing beer some eleven thousand years ago. Lately, however, this team has been more noncommittal, referring simply to "cooking of cereals." Still, whatever the case at Göbekli Tepe, once durable pottery had arrived on the scene it seems that everybody was busy fermenting whatever was available, using a variety of independently conceived approaches.

The earliest evidence we have for deliberate fermentation comes in the form of chemical residues that were absorbed into the pottery fabrics of vessels found at the early Neolithic site of Jiahu in northern China. There, at around nine thousand years ago or very shortly thereafter, residents made a fermented "grog," or mixed beverage, from honey, hawthorn fruit and/or grapes, and rice (thus combining the ingredients of a beer, a mead, and a wine). A fermented beverage recently reconstituted from these natural products (and marketed as a beer) came in at a respectable (and intentionally targeted) 9 percent ABV (alcohol by volume). Farther west, the earliest fermented beverage yet documented (in the form of residues left behind in pottery jars) is a grape wine from the Republic of Georgia that was made around eight thousand years ago. And to the south, in Mesopotamia, by about five thousand years ago the very early Sumerian civilization was not only awash with beer, but also ruled by an elite class that flaunted grape wines as a symbol of authority.

It is clear that people all over the world were quick to put their existing enzymatic prowess to the test just as soon as they had the technology to do so. And however we may choose to rationalize or even dismiss this evident fact, humans everywhere are plainly attracted to alcohol, even as they are often repelled by the consequences of consuming it to excess. This strongly implies that even if the regular consumption of alcoholic beverages appeared relatively late in the history of the human species, that was only because, at first, we did not have the technologies necessary to produce and store alcohol in quantity. The predilection for alcohol was always there, lurking in the background until the conditions were right for expressing it. This observation is hardly a surprise. For not only is alcohol consumption a feature of human cultures throughout the world, but other creatures enjoy a tipple, too.

The most famous of these other creatures, at least among scientists, is the pen-tailed treeshrew of Malaysia. This tiny (two-ounce) mammal is renowned for its ability to binge off the naturally fermenting nectar of the bertam palm for hours at a time, while apparently enjoying the experience for much more than its purely nutritive value. Since the nectar ferments to an ABV of almost 4 percent—about the same as a traditional English beer—this long indulgence means that the tiny animal may easily consume the human equivalent of three bottles of wine in a single session. Yet it does so with no sign of inebriation—which is lucky, because dangerous predators abound in its habitat and any relaxation of vigilance could be fatal. Nobody knows exactly how the pen-tail pulls off this remarkable trick; its ADH enzymes have yet to be characterized, although a close relative has been tested and found to possess the common form.

Almost equally famous for its love of naturally fermented alcohol is our fellow primate the Panamanian howler monkey—although, alas, this closer relative is indeed prone to inebriation. A quarter of a century ago, researchers noticed a howler feeding with unusual enthusiasm on the fruits of the *Astrocaryum* palm. The individual became so frenzied that the observers suspected he might be drunk, a state confirmed by analysis of the partly consumed fruit he kept dropping to the forest floor. Initial calculations suggested that the twenty-pound monkey had consumed the equivalent of ten bar drinks in a single binge.

This observation made the biologist Robert Dudley wonder why the monkey was so fond of alcohol. His conclusion: the answer lay less in the alcohol itself than in the highly nutritious sugar that was being naturally fermented to produce it. Howler monkeys are principally fruit eaters, and (like humans) they come from an essentially fruit-eating ancestry. They actively need ripe fruit, and can deal with a little alcohol (indeed, primates share some so-called inebriation genes with rel-

Figure 1.2. The aye-aye, the most specialized of the lemurs of Madagascar, and the only primate other than humans and the African apes to possess the ethanol-active version of ADH4.

atives as remote as fruit flies). What's more, while the presence of desirable sugars in any nonfermented ripe fruit has to be indirectly inferred from its color, fermentation sends out strong fumes that can be detected at a distance by a keen nose. Consequently, if a frugivore is guided by scent to a ripe and incidentally (but rewardingly) alcoholic fruit, its diet will automatically be improved. This is an obvious advantage for the individual and, on a larger scale, for its species. Hence Dudley's "drunken monkey hypothesis" that views our own human predilection for alcohol as, ahem, an "evolutionary hangover." But if there is anything causative in the ADH4 story—and even if not—one has to wonder why more frugivores did not acquire the new version of the enzyme. After all, the mutation concerned is a simple one, and we know it occurred more than once among the primates, since Madagascar's strangely beautiful aye-aye also boasts the modified enzyme (fig. 1.2).

Closer to home, our very close relative the chimpanzee shares our improved ADH4 with us, so it would perhaps be surprising if members of its genus did not occasionally capitalize on that fact. And indeed, it turns out that chimpanzees occasionally do indulge in a quick tipple. Researchers at a site called Bossou in the West African country of Guinea have reported seeing chimpanzees emerge repeatedly from the somewhat degraded forest in which they live, heading for a nearby plantation of raffia palms. Not much for them there, you might think. But it turned out that plantation workers were tapping the raffia trees to obtain sugar-rich sap that dripped into plastic containers. Inside those containers the sap rapidly and spontaneously fermented into an alcoholic palm toddy which, depending on time elapsed, would achieve between 3 and 7 percent ABV. While the workers' attention was elsewhere, the chimpanzees would descend on the plantation and scoop the alcoholic liquid from the containers, cleverly using crumpled leaves as "sponges" to soak it up. In our experience, palm toddy starts off as a sweet and delicately flavored beverage, but by the time it reaches the alcohol levels at which the chimpanzees consumed it, it is generally very pungent and—to us—little short of repellent. But wherever it was in its taste evolution, the chimpanzees evidently loved the stuff, dipping and emptying their sponges on average every few seconds, for minutes on end. It is hard to know if the chimpanzees liked the toddy because of its gustatory qualities or because of the buzz they undoubtedly got out of it (observers reported "behavioural signs of inebriation"); but, based somewhat unscientifically on our personal opinion of excessively fermented palm toddy, we're guessing the buzz played a significant role.

Anthropologists Ruth Thomsen and Anja Zschoke, however, wanted to find out more. After reading about drunken monkeys, ADH4, and the Bossou chimpanzees, they set up an experiment. For ten days, they offered some zoo chimpanzees daily treats consisting of apple purée either with or without rum flavoring. After showing some initial interest in the rum-flavored samples, the chimpanzees lost any preference, leading the researchers to conclude that they did "not prefer alcohol-rich fruits more than non-alcoholic fruits," and thereby to reject the drunken monkey idea. But looking a bit closer, we found that the apple purée was not flavored with actual rum. Doubtless because of some footling bureaucratic requirement, the substance used was a rum flavoring rather than the real thing, and the alcohol level in the purée was a mere 0.5 percent ABV. In other words, the most we can conclude from the experiment is that chimpanzees don't care much for nonalcoholic rum substitutes. We have no idea whether a good dose of real, 80-proof rum in the apple purée would have changed the result, but we like to think that the Bossou chimps were going for the buzz.

Whether or not chimpanzees actually like alcohol, it is clear that we humans are not alone in our attraction to alcoholic beverages. But we are unique in our dual ability both to produce alcohol at will, and to consume it in relatively large quantities. And it seems that, from the beginning, considerable ingenuity and effort went into the human production of booze. The earliest producers of alcoholic drinks fermented whatever suitable substances they could lay their hands on, clearly aiming for the alcohol itself as well as for flavor: the categories of beer, wine, and mead were sorted out only later, according to tradition and the principal resources at hand. Distillation came along pretty late in the history of human alcohol use, almost certainly due to the fancy additional technology required. But once the new approach was on the scene, its manifold possibilities were assiduously and enthusiastically explored, with the result that today every bar features a vast range of distilled beverages for clients to try.

The world of alcoholic drinks is a pretty ecumenical one. There is room behind all but the most specialist bars for a full complement of beer, wine, and spirits (and nowadays, even mead again), suggesting that it is part of the human psyche to wish to enjoy alcohol in all of its expressions. What is more, most drinkers recognize that appropriate occasions exist for consuming all of these kinds of drinks, so that instead of being in competition, they are, in the main, complementary. After all, while you would not want to slake your thirst with grappa on a hot day, chances are you wouldn't wish to accompany your tiramisu with a sour ale. Alcohol's most wonderful property is its ability to deliver a wide array of textures, flavors, and aromas, and so to satisfy the human craving for sheer gustatory variety.

2

A Brief History of Distilling

Before tackling the history of distillation, we felt we had to sample a product from Europe's oldest more-or-less continuously operating licensed distiller, located near Ireland's rugged, windswept northwest coast. Granted its initial license to distill in 1608, even before the Bushmills Old Distillery Company was formed in 1784, the current Bushmills distillery has long carried the banner for Irish whiskeys, triple-distilling its creations in rows of traditional squat, long-necked copper stills. This was the sixteen-year-old bottling, made from a bill of 100 percent malted barley. Its components had spent their first fifteen years in casks previously used either for Oloroso sherries or for American bourbons, before being assembled into large port "pipes" for nine months of finishing. The end result reflected this complex history. Golden in the glass with reddish highlights, this whiskey presented on the nose as fragrant and distinctly fruity. It gave our palates toffee and dark fruits as well as hints of sherry, all superimposed on the rich malty backbone contributed by the traditional mash and those gleaming, old-fashioned pot stills. We could taste every part of this whiskey's storied journey, from mash to glass.

Next time you prepare some salty soup or stew on the stove, put a transparent cover over it and let it come to a boil. After a few minutes, you will see some condensate form under the lid. Carefully remove the lid, tilting it to collect some of the drips. Taste those drips, and then taste some of the stew from which it was derived. There should be a big difference. The condensate should taste neutral in contrast to the salty stew.

What you have just experienced is a simple form of distillation, a process that uses heat to vaporize, and so to separate, different compounds out of a liquid mixture at different temperatures. For example, ethanol, the desirable form of alcohol, boils at 173.1 degrees Fahrenheit, while water boils at 212 degrees. So, if you heat up a liquid—a wine, say—that contains both to an intermediate temperature, the alcohol will boil off and can be condensed and collected separately, while the water will stay in its receptacle. But beware: poisonous methanol will vaporize before the drinkable ethanol does, so distillers must discard the "heads" that come out of the still first (as well as the "tails" that come out at the end; the sweet spot for any particular application is somewhere in the middle). Because the alcoholic "wash" you start with is full of compounds other than methanol and ethanol that may or may not be desirable in the final product, and that also have their own boiling points, modern distillers use a technique called fractionation to tweak the target temperature and achieve the desired result. It is not known exactly when and where the basic principle of fractionation was discovered and put to practical use, but fortunately, that has not stopped historians from speculating.

On April 27, 1906, in an address to the annual meeting of the United Kingdom's Institute of Brewing and Distilling, the food scientist Thomas Fairley presented a stimulating discussion of the early history of distillation. His exposition was well received by the audience at the Queen's Hotel in Leeds, and it generated a great deal of discussion. In framing his talk, Fairley leaned heavily on a pamphlet written in 1901 by the soil scientist Oswald Schreiner. Fortunately for us, Fairley's address was published in the *Journal of the Institute of Brewing,* and Schreiner's pamphlet has also survived. Both publications are rich with illustrations; and although both are written in the quaint language of the nineteenth-century gentleman-scientist, they contribute several important ideas about the early history of distillation. These include speculation on where the first distillation apparatus was invented, observations on the shapes of early stills, and some suppositions about the origins of alcoholic beverages. Perhaps the most insightful bit of speculation in these works is Fairley's remark that "a knowledge of the art of distillation might arise in different countries independently." Fairley recognized that many cultures had boiled food and other substances since antiquity, and that it wouldn't have been a big leap to realize something significant was happening when steam was released. To Fairley, the invention of stills and distillation was just a few small steps away from this realization, which might have dawned on people in various times and places.

To support his claim, Fairley points to the variety of distillation apparatuses that have existed, and are used to this day, in cultures around the world. In his discussion he ranges from ancient Greece and Rome to Egypt, Arabia, Europe, India, Ceylon, Japan, China, Tibet, Bhutan, the Caucasus, Tahiti, and Peru, drawing from accounts by such historical luminaries as the Greek Aristotle, the Roman Pliny the Younger, the Alexandrian Zosimos, and the Arab Geber (Jābir ibn Hayyān). Fairley also notes how, from time immemorial, sailors have obtained potable water by boiling salty sea water and collecting the condensate. Aristotle himself had been aware of this, writing in his *Meteorology* that "sea water can be made drinkable by vaporization; other liquids behave in the same way."

As for the origin of the word *distillation,* Fairley explains that neither the Greeks nor the Romans, who between them gave us the roots of most of our modern scientific and technical terms, had a word that held this exact meaning. In fact, the only way to describe a distillate in Latin would be via some circumlocution such as *rei succum subjectis ignibus exprimare* (the expressed juice of the matter subject to the fires). Nonetheless, like most scientific terms, the English word *distillation* does have a Latin root, coming from the Latin noun *stilla* (a drop), via the Latin extension *de-stillo,* later simplified to *distillo* and finally modified into the modern English form. Given this origin, it is not surprising that in the early days the notion of distillation had a more generalized meaning than it has today, referring to the separation of the light part of any fluid from the heavier stuff via the process of drop formation. It was only in the fourteenth century that the term began to be associated explicitly with alcohol production.

Back then, of course, nobody knew what alcohol was, merely recognizing that the clear liquid resulting from distillation had distinctive effects when ingested: right up through the eighteenth century, distillates were simply thought to be alternate forms of water. Given the knowledge of the time, that was entirely reasonable: any clear liquid derived from the process of heating and condensing would have looked just like water to people who hadn't taken college chemistry. Indeed, in his paean to spirits, Pliny the Younger said, "Oh! Wondrous craft of vices! By some mode or other it was discovered that water itself might be made to inebriate." It is no surprise, then, that early references to alcohol typically call it a form of water, as in *aqua vitae* (Latin), *eau-de-vie* (French), and even the Celtic *uisge-beatha,* all of which mean "water of life."

The word *alcohol* itself was said by Fairley to originate from the Semitic word *kuhl* (also *kohol* or *koh'l*), which was in prominent use in both the Arab and Hebrew languages. *Kohol* referred to a fine powder obtained by the distillation of an ore, and it eventually came to stand for any compound or substance that is fine or pure. With the Arabic article *al* added (*al-kuhl*), the word *alcohol* was born. As we will see, European usage was initially associated with the early meaning of *distillation,* which, according to the chemist Norbert Kockmann, referred to any separation operation that included "filtration, crystallization, extraction, sublimation, or mechanical pressing of oil." That's a pretty broad category. Perhaps the most engaging part of Fairley's address involved his explanations for the names given to the variously shaped stills that have been created. Because those shapes resembled animals, he established a veritable zoo of distillation flasks (Figure 2.1). Hence, we have the Struthio (named after the ostrich—*Struthio camelus*) or the testudo (named after the turtle genus *Testudo*), and the multiheaded apparatus called the hydra, after the Greek mythical beast.

It is difficult to pin down exactly when the first still was invented, largely because devices of this kind are found in so many cultures; and except for those found in Western Europe we have very little information on the age of distillation processes used today. But we do know for sure that distilling came along after

Figure 2.1. Still shapes named after animals. *Left to right, top row:* Struthio, testudo, ursus, pellicanus; *lower row:* hydra, serpentine, and ciconia. After Fairley (1907).

beer and wine making, because it is from those lower-alcohol-content beverages that spirits are usually distilled. This observation narrows down the origin of spirits distillation to sometime after 6000 BCE or so. But that still leaves a substantial time gap before we have specific records that anyone was actually distilling. It didn't help that in 296 CE the Roman emperor Diocletian, fearing that any success by alchemists in their gold-producing endeavors would debase his coinage, declared war on these people, the first recorded distillers, and had their manuscripts burned. Indeed, most of our knowledge about the early existence of stills in global cultures dates from as recently as the century of exploration between 1500 and 1600 CE, though the explorers' records are supplemented by some much earlier archaeological indicators as well.

Based on surviving historical evidence, some scholars claim that Persia was the birthplace of the still, pointing to well-established practices of distilling rose water and rose oils in this region. Other historians believe that the source was India, or maybe Egypt. Perhaps more suggestive evidence comes from archaeology, which points toward the Fertile Crescent in the Near East. A 5,500-year-old Sumerian site in Mesopotamia has yielded incomplete remains of what may have been evaporation vessels belonging to devices used to distill substances like essential oils, rose water, and wood turpentine (Figure 2.2). Though such rudimentary stills were not very efficient, once the basic condensation design had taken hold, improvements soon followed. Most importantly, outlet pipes were run from

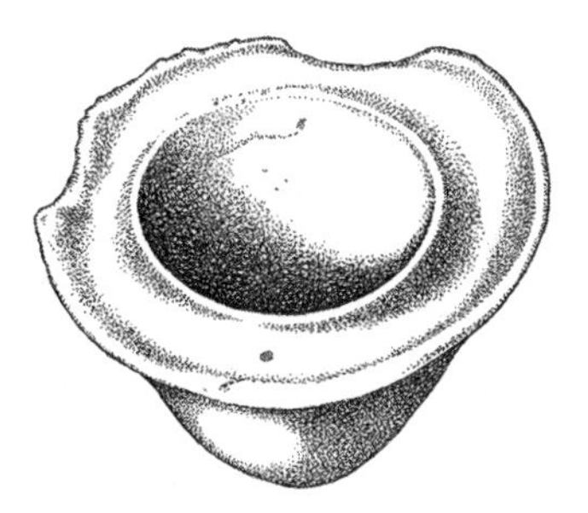

Figure 2.2. What might be a condensation vessel from an archaeological site near Mosul, Iraq. The device lacks the cooling lid which would have been placed atop it. The concave rim around the top is the collection part of the still. Condensation droplets from a boiling solution in this device would have collected on the bottom side of the lid, which was appropriately tapered. The drops would then have slid down the lid into the collection area of the still. The boiling pot is about 50 cm in diameter. After Kockmann (2014).

the collection trough to a collector vessel on the outside of the device, and better cooling was achieved by moving the collection trough to the cooler lid, or by actively cooling the collection device with wet sand or water jackets. We can view these advances as part of a genealogical process in which the great-great-grandchildren of these innovations became what are broadly known today as "pot stills."

Following Kockmann's lead, we can divide the history of distillation and the development of stills in western Eurasia into roughly five major periods: the preclassical to classical period (6000 BCE to 700 CE), the Arabian-Alembic period (700 to 1450 CE), the Renaissance-science period (1450 to 1800 CE), the period of industrialization (1800 to 1950 CE), and the modern period of integration. We have already taken note of the sparse evidence remaining from the Neolithic to classical periods; let's now look at the four later periods of distillation history.

The transition to the Arabian-Alembic period is marked by the widespread introduction of the alembic-style pot still. This device, according to the fourth-century alchemist Zosimos of Panopolis, had been devised a century or two earlier in post-Alexandrian Egypt by an enigmatic figure known as Maria the Jewess, who was possibly also the mother of alchemy (the word alchemy comes from the Arabic *al-Kemet,* or "Egypt"). Zosimos described the still illustrated in Figure 2.3, a style that comes from early first-millennium CE Greece and is variously known as the Still of Democritus, or the Alembic of Synesios. It has four components, and its design served as the basic model for alchemical stills well into the second millennium. The *alembic* (helmet) on top is connected via a pipe to the *receptaculum* (receiving vessel) below, and the heating apparatus (called the *cucurbita,*

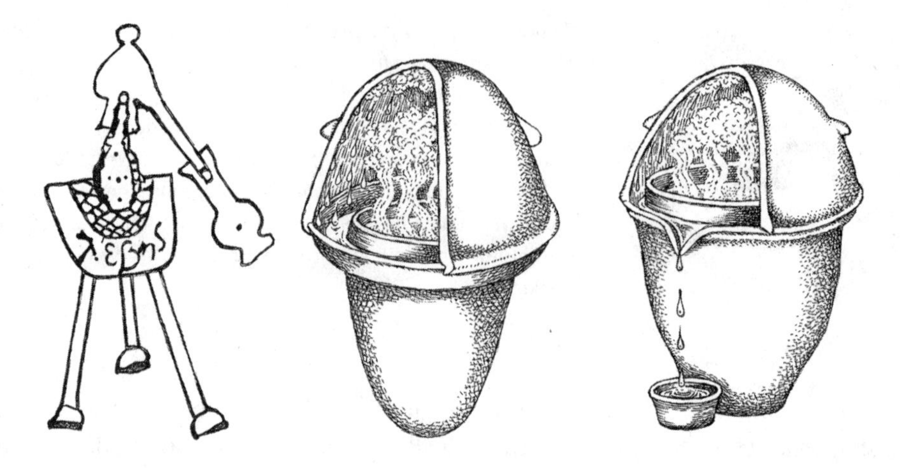

Figure 2.3. *Left:* Contemporary diagram of the Still of Democritus (Alembic of Synesios), early first millennium CE. Note the alembic architecture of the still and the inscription on the heating vessel—λεβης—which is Greek for "kettle" (*lebes*). *Center and right:* Cutaway views of Iranian stills. After Kockmann (2014).

because it was shaped like a gourd) is labeled in Greek characters and supported by a tripod.

Simple stills of this general kind were widely employed by Arab alchemists from the eighth century and were commonly used right up to the Renaissance. The word *alembic,* now used for a distilling device consisting of two separate vessels connected by a tube, ultimately stems from the Greek root *ambix,* meaning cup. This etymology would seem at first glance to support the idea that the device originated in post-Hellenistic Egypt, but the term *alembic* itself actually comes down to us via the Arabic word derived from it, *al-inbiq,* indicating the route by which its name arrived in modern European languages. Those early pot stills are also referred to as helmet stills, and were more than likely invented by alchemists.

Arabian influences in the Arabian-Alembic period are well documented from the eighth century onward, as Arab science flourished and many people in the Near East became well versed in chemical and alchemical processes. We have already mentioned Geber (Jābir ibn Hayyān, the inventor of algebra—*al-Jabr*), who was also the leading chemist of his time. He noted that the products of distillation were flammable, and recommended that distillation be accomplished using glass vessels because of the material's nonreactive nature. Some of the alchemical writings attributed to Geber were translated into Latin, and the knowledge they contained was duly transferred to European cultures (see Chapter 3). Significantly, while the Arabs were evidently more interested in distillation as a means of creating essential oils and other nonalcoholic essences, once the alcohol-loving Europeans had learned distillation they soon began using the technology to convert wines and other fermented beverages into high-alcohol spirits, following a sort of "less is more" attitude.

Early distillation in Europe was usually undertaken by alchemists, and at first the distilled products were used mostly as medicines (an application that was apparently given a boost during the Black Death of 1347–1350, when the alcoholic beverages didn't cure, but were the best palliative available). The science of distillation reached a new level of sophistication when, early in the fourteenth century, alchemists refined two distinct approaches that were highly dependent on the shape of the still: *ad ascensum* and *ad descensum*. The *ad ascensum* approach co-opted rising vapors to form the distillate, while *ad descensum* employed downward vapor flow and was eventually largely discarded. Another development in the early Renaissance-science period was the invention of the dephlegmator, a small receptacle attached to the pipe between the cooling cap and the collection vessel. This ingenious device was designed to capture and separate out components of the distillate that the traditional procedure had allowed into the final solution. As the condensate ran out of the heating vessel via the cooling cap, the less volatile fractions would precipitate and drop out before the more volatile ones like alcohol did. Alchemists created this bypass point because those less-volatile compounds were of interest to their work in alchemy, but drinkers were thrilled, because they wanted those compounds out of their grog. The dephlegmator was a win-win in all possible ways.

Most distillation today descends from the Arab-European tradition, but that doesn't mean that the potential for concentrating alcohol was not also being tried out in other parts of the world. Timescales are hazy, but sometime prior to Spanish colonization, people in the Philippines were probably already making the spirit known nowadays as lambanóg. This liquor is made from coconut or nipa palm sap that is obtained in much the same way as the other palm toddies (*tubas*) that are common in the tropical world (and beloved by the Bossou chimpanzees). Historically, lambanóg was produced by the double distillation of sap that had fermented for two days on hearth-driven stills made up of two pans, or wood barrels, connected by a hollowed log. That double-distillation process more than likely resulted from incorporating Spanish distillation techniques into older local methods, and it was certainly being practiced at least as early as 1574. Crude as the equipment might appear, its use resulted in a smooth, vodka-like spirit with very high alcohol content of 40 to 45 percent ABV on first distillation, and above 80 percent on the second. Although aficionados prize the purity of lambanóg, modern versions made on more up-to-date equipment are sometimes marketed in flavors ranging from cinnamon to bubblegum.

China's deep history of distillation is better documented. Distilled spirits were already being made in that ancient land by the time Marco Polo arrived in the thirteenth century, as confirmed by the remains of a twelfth-century still at Qinglong, in northeastern Hebei Province. Whether distilling was an independent indigenous development, or whether the practice came from the west along the Silk Road, is not yet entirely clear, though most authorities believe that there was a connection. Whatever the case, it is certainly appropriate that one of the oldest operating distilleries in the world is found in China, at Luzhou, in southern Szechuan. Using the techniques described in Chapter 17, this venerable establishment has been producing baijiu since the mid-sixteenth century.

Another closely related Eastern distillation tradition comes from Mongolia. As the result of Russian influence there are currently hundreds of vodka distilleries in Mongolia, but traditional Mongolian spirits differ greatly from vodka. Fermented mare's milk or *airag* (*kumiss* in its Turkic rendition) is the most notorious Mongolian alcoholic beverage. It is not a product of distillation, and at 2 to 3 percent ABV it is mild compared to other Mongolian alcohols. But it can be made stronger through the kind of freeze distillation believed to have produced the earliest Russian vodkas (see Chapter 10). It can also be converted into *arkhi,* a traditional Mongolian spirit that is widely produced in Mongolian households, although mostly from the fermented cow's milk yogurt known as *kefir.* To make arkhi, the kefir (or airag) is placed into a concave wok on a stove. A bowl-shaped collection device is then placed over the wok, supported by a cylindrical collar and enclosed on top by a concave lid filled with cold water. When heated, the kefir releases its alcohol, which condenses onto the cold lid and drips into the central bowl. After two passes, the collected condensate is ready to drink as a clear beverage of about 10 percent ABV. The taste of the arkhi at this stage is a bit rancid for those not accustomed to it, and further passes to raise the alcohol content only accentuate that flavor profile.

Returning to Europe, the major shift from the Arabian-Alembic period to the Renaissance-science period has been attributed by Kockmann to the invention of the printed book. Movable print is one of the major factors that drove science and culture generally in the Renaissance, and it allowed for the wider distribution of distillation recipes and procedures that until around 1450 CE had been restricted in Europe to monasteries, medical schools, and apothecaries. Using the new communication technology, several authors from the late fifteenth to the mid-sixteenth centuries spread knowledge of distillation far and wide.

Distillation experts of this era also wrote about important concepts of brewing that were under development. Most important, the best materials for making the distillation vessels were explored extensively, with glass, clay, and other ceramics being used as materials for both alembics and collection vessels. Copper and lead were also used for both these purposes, but alchemists initially steered away from using these metals because they reacted with the chemicals they contained. Alcohol distilled from a lead vessel would, for example, turn milky white, alerting the alchemist to the fact that something untoward was happening. The resulting toxic material was fortunately rejected. Copper, in contrast, was soon discovered to remove the sulfuric compounds that formed in many distillates. Because sulfur is the distiller's curse, producing unpleasant vegetal and "rotten egg" aromas in the product, copper has been favored as a material for alcohol stills ever since.

By the end of the fifteenth century the Rosenhut "hooded still" had been developed, mainly for the isolation of medical liquids. This kind of still was an ingenious all-in-one structure, complete with a fire area for heating the material, a built-in conical heating vessel, a connected cooling cap, and an alembic arm (Figure 2.4).

The use of steam rather than direct fire to warm the heating vessel was another innovation of the early Renaissance period. Kockmann points to the sixteenth-century French astrologer Claude Dariot as the first European to use steam distillation, an important advance because it allowed the alchemists precise control of the temperature of the material being distilled. The technique involved boiling water to produce steam that would rise to gently heat the vessel holding the material to be distilled (usually a cucurbita, as in Figure 2.3). The

Figure 2.4. A Rosenhut still. After Wikimedia Commons. https://de.wikipedia.org/wiki/Datei:Rosenhut.png.

rest of the distillation then proceeded as usual. Other means of controlling temperature in the distillation process were also developed in this golden age of still-making, most of them involving improved heat transfer.

The steam method demonstrated how incredibly important temperature control is in any distillation procedure, during both cooling and heating. Many alchemists had simply used the air around the still to cool the distillate, and some distillation apparatuses still use this age-old method of cooling. But another early method was to use a water jacket, which involved passing the exit pipe through a bath of cold water that could be changed when the heat transfer became too intense. The next advance was to have cold water constantly running through the water jacket around the exit arm of the still. Either way, the water jacket not only allowed for better control of the cooling phase, but also reduced the chances that the heating vessel and exit tube would be cracked by intense heat. A related advance was the invention of coiled cooling pipes, often known as "worms." The coiled pipe lengthened the route through which the distillate had to pass on its way to condensation, resulting in more efficient cooling, especially after the coils had been incorporated into the water jacket.

Equipped with the advances made during the Renaissance, later alchemists had a well-tuned device with which to ply their trade. But a funny thing happened on the way to the Industrial Period: science flourished, while alchemy faded away. This transition ultimately led to the creation of three different kinds of stills: laboratory stills designed to fill the needs of rigorous science; industrial stills for mass-producing compounds important for commerce, such as the sulfuric acid required to dye the linen and cotton fabrics that became popular during the Industrial Revolution; and stills for producing medicinal and recreational alcohol, already a thriving business during the late medieval period.

One good place to begin the story of distilled beverages is the year 1500, when Hieronymus Brunschwig's book *Liber de arte distillandi,* one of the earliest printed works on spirits distilling, was published. Fun fact: when it was translated into English in 1527, this book's simple title, *Book of the Art of Distilling,* became *The Virtuous Book of Distilling.* Why virtuous? Well, that had everything to do with the view of distillates as a gift from God, for Brunschwig banged the drum for spirits as a miracle cure for almost every disease imaginable. As his book's English translator, Laurence Andrew, rather awkwardly declared in his preface: "Behold how much it exceeds to use medicine of efficacy natural by God ordained than wicked words or charms of efficacy unnatural by the devil invented." In other words, at the turn of the sixteenth century the products of distillation ("medicine of efficacy natural") were seen as gifts given by God to cure many of the ills of the day.

What a difference a century makes! Contrast Andrew's note of solemn veneration to the tone adopted in 1618 by his compatriot John Taylor, in his *Pennyless Pilgrimage.* Taylor's carefree subtitle says it all: *How He Traveled from London to Edenborough in Scotland, Not Carrying Any Money To or Fro, Neither Begging, Borrowing, or Asking Meate, Drinke or Lodging.* This subtitle prepares us for such immortal lines as:

> And I entreat you take these words for no-lies,
> I had good Aqua vitæ, Rosa so-lies:
> With sweet Ambrosia, (the gods' own drink)
> Most excellent gear for mortals, as I think.

Taylor, self-described as the "Kings Majesties water-poet" (watermen of the time were notorious drunkards) clearly enjoyed some strongly alcoholic beverages as well as vast quantities of ale during his jaunt from London to Edinburgh, and certainly not in pursuit of a healthy mind in a healthy body. In less than a century, there had evidently been a big shift among Europeans in the use of the products of distillation, all the way from alchemical/medical purposes, to applications that advanced the development of modern chemistry and chemical engineering, to the creation of inebriating beverages that the population drank for fun. Perhaps it is not entirely coincidental that this was also the time that Europeans began to distill sugarcane from the New World to make alcohol (see Chapter 14).

When the impecunious John Taylor made his trek to Edinburgh, the spirits he drank were the products of alembics that Zosimos of Panopolis might have recognized. That was because one major technological innovation was still needed before alcoholic spirits could truly be mass-produced. As it was, traditional pot stills produced alcohol in batches. A given amount of wash would be vaporized and condensed, and then the still would have to be emptied and cleaned before the process could start again. What is more, even the most efficient modern pot still (hugely more effective than in Taylor's time) will not produce a spirit of more than 30 to 35 percent alcohol by volume (ABV) from a beginning wash of 12 percent alcohol or thereabouts. To produce a stronger spirit, the distillate must be passed through the still multiple times, in a process that not only raises the alcohol to the desired level, but also ensures the elimination of many of the (wanted or unwanted) compounds that were present in the initial wash. Pot stills can sometimes be pretty large (one monster constructed early in the nineteenth century held 143,000 liters), but using them is nonetheless a time-consuming and labor-intensive business, and the purity of the resulting spirit is limited. Spirits that could be manufactured on a truly industrial scale thus required something more efficient. Enter the column still.

Both an Irishman and a Scotsman are credited with the invention of the column still. But this epochal event did not happen before several ingenious chemists and distillers, notably in France, had added some important innovations to distillation procedures in the late eighteenth and early nineteenth centuries. Some of the devices invented in this period even approximated column stills. But due to the vagaries of the patents process, credit for the column still that we continue to use today is shared by the Scotsman Robert Stein and the Irishman Aeneas Coffey. Stein initially conceived the column still, patenting it in 1828. His design was a huge advance both over the pot stills we have so far encountered, and over the

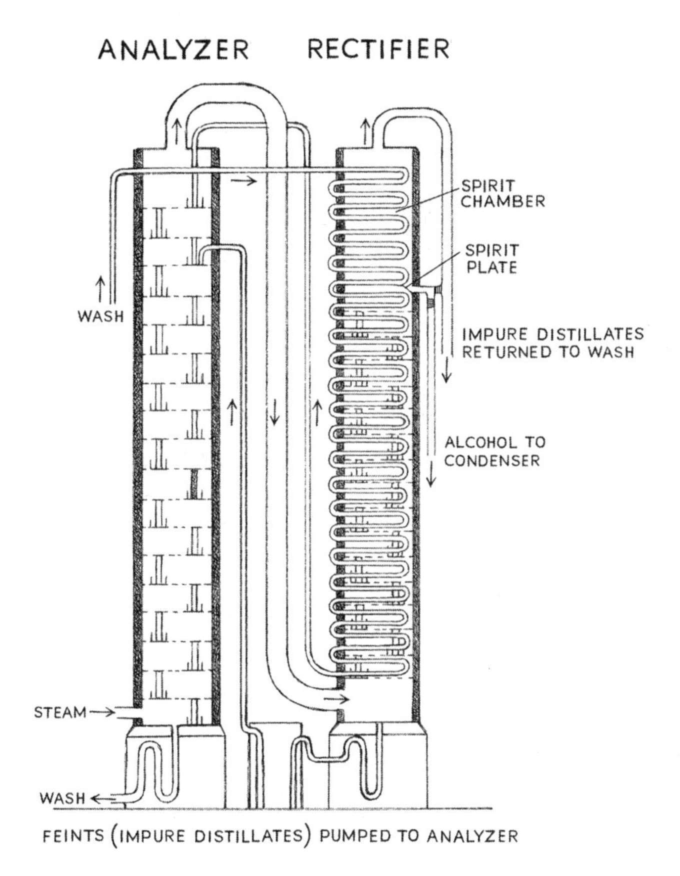

Figure 2.5. Schematic of an early two-column still.

mostly incrementally improved versions that had come into use during the decades preceding his patent application. Stills made to Stein's specifications, however, required frequent cleaning and were limited in the purity of the spirit they produced. Coffey took Stein's concept and ran with it, patenting his own improved version in 1830 (Figure 2.5). The result is now known as the "Coffey" or "continuous" still.

Continuous stills work on the principle of fractional distillation. Pot stills produce a distillate that contains many kinds of molecules ("congeners") along with the ethanol. But Coffey stills conserve only a fraction of what is being distilled, producing high-grade alcohol, usually 95 to 96 percent pure (above this level, water and ethanol form an azeotrope, meaning that they assume the same

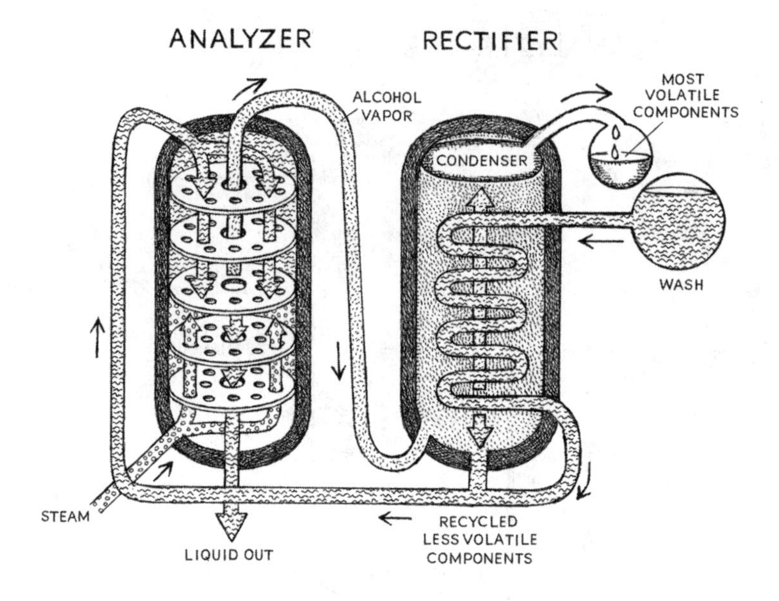

Figure 2.6. Diagram of idealized Coffey two-column still. Both columns are preheated by steam. After Wikipedia Commons.

boiling point). The end products of continuous stills are thus typically diluted before they are bottled at around 40 percent ABV.

The big physical contrast between the pot still and the Coffey still is the tall two-distilling-column setup in the Coffey still, versus the single and usually bulbous distilling vessel of the pot still. The Coffey still has two long vertical columns, the first called the "analyzer" and the second the "rectifier," both heated by steam (Figure 2.6). Steam is allowed to rise in the analyzer at the same time as a wash descends the column through a series of levels at which perforated plates are placed. The alcohol from the wash is passed through to the rectifier, where it is circulated, condensed, and collected. The neat thing about the Coffey still is that the starting material can be cycled endlessly through the two columns, so that production can continue uninterrupted.

The Coffey still enjoyed immediate popularity among spirit producers, and was adopted almost everywhere—except, ironically, in the inventor's native country, home at the time to the world's biggest whiskey industry. Most Irish distillers felt that, despite the higher alcohol content it delivered, the Coffey device was deficient when it came to the taste of the distillate. They had a point, because continuous distilling efficiently removes most of the contaminants and trace elements present in the original wash, and so eliminates more of the congeners that give the flavor and bite that make Irish whiskeys so cherished by drinkers around the world.

Although Coffey's compatriots lacked enthusiasm for his continuous still, Stein's home country of Scotland gladly adopted his invention, using it to make the Scotch whiskies that were mainly destined for blending. Ultimately, this led to the divorce between *whisky* and *whiskey* drinkers, as recounted in Chapter 3. Some American and Irish whiskey-makers eventually adopted the Coffey still, and today both pot stills and continuous stills are used in those industries—although worldwide, many more spirits are produced on continuous stills than on alembics. Although it is hard to generalize, it is reasonably fair to say that today most "craft" distilling is done in alembics, while spirits produced in industrial quantities come from continuous stills. But not always: even the largest Cognac houses are obliged by law to use copper pot stills, while many craft distillers, notably of gins and vodkas, begin with "neutral" (high purity) spirits produced in gigantic continuous columns. As a result, most spirits categories on the market today may be made on either kind of still; and there are now "hybrid" stills that make the original definitions and rules all the more confusing.

Because of the relative inefficiency of the process, most spirits made in alembics are distilled more than once, with twice being obligatory for Cognacs (though Armagnacs get only a single pass), and ten times or more are sometimes claimed. But be careful when you see any spirit made in a continuous still that claims to be multiply distilled (as many vodkas do). That claim is basically meaningless, and likely to be advertising hype. A tall column may have many dozens of those perforated plates: how many "distillations" does that imply?

3

Spirits, History, and Culture

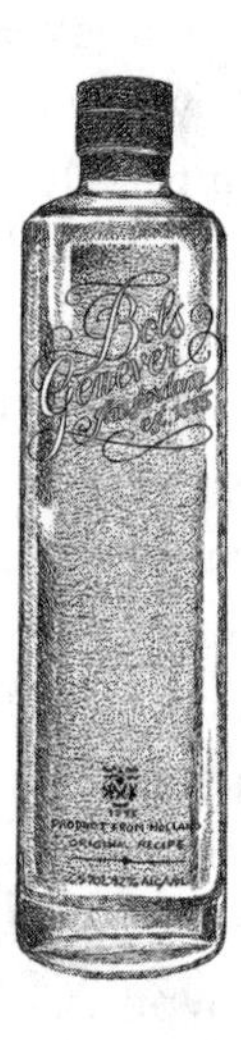

People have been making spirits for a long time, and doubtless nothing available on the market today tastes much like the liquid that dribbled off the very first still. But how about a spirit made by Europe's oldest surviving distilling company, according to a two-hundred-year-old recipe? Presented in a tall, slender bottle, this unaged genever had the traditional Dutch malt bill of rye, corn, and wheat. A direct descendant of a spirit that was first shipped to America in the early nineteenth century, and that subsequently helped spark the cocktail revolution, it had been triple-distilled in copper pot stills before being infused with a separate botanical distillate. Unctuously clear in the glass, it gave off a powerful nose of juniper berries, with a trace of freshly crushed mint over a strong malty base. These dominant flavors, gently warmed by the prominent alcohol, lingered on the palate and were accompanied by a hint of crunchy celery in the finish. We were happy to drink this delightful, complex spirit entirely on its own. But then a small twist of lemon zest transformed it into an altogether brighter and more expansive experience—and we understood immediately why nineteenth-century American bartenders had latched onto its potential to add kick and personality to a whole range of classic cocktails.

Knowledge of the principles of distillation may extend back to classical times in the West, and to Han times in eastern Asia. But more important from the spirits drinker's point of view was when those principles were used in the widespread production of recreational beverages, a breakthrough that was probably achieved in Europe only in the fourteenth century. This relative recency means, of course, that the cultural roots of spirits drinking are historically a lot shallower than those of wine or beer consumption. Still, after half a millennium of recreational distillation, those roots have become no less deeply and widely entrenched in local human behavior; for beer, wine, and spirits, while all containing alcohol, play roles that are much more complementary than competitive in the human experience.

By the time distillation developed, societies worldwide had long been drinking beer, wine, and other fermented beverages. Sixteenth-century Spanish conquistadors rapidly discovered that the consumption of alcohol was entrenched throughout the southern New World from Mexico to Chile, with a bewildering variety of social drinking customs to match. The Inca Empire centered in Peru was, for example, fueled not only by the mildly narcotic coca leaf that sustained peasants laboring at high altitudes, but also by the consumption of chicha, a brew that was usually made by fermenting chewed or malted corn. Chicha held a central place in Inca ceremonies undertaken to celebrate the harvest and to propitiate the gods, and it was sometimes force-fed to human captives awaiting sacrifice. Even the hard-bitten Spanish conquistadors were appalled by the large volumes consumed in the wake of such ceremonies. A few years earlier, their compatriots who had landed farther north had been equally amazed by the huge variety of rules that governed the consumption of pulque (fermented sap of the agave plant) among the relatively abstemious Mesoamerican Aztecs. As usual, those rules revealed a conflicted attitude toward alcohol: nobility excepted, only adults over fifty-two years old were allowed to drink pulque at will, and the populace in general was sternly warned about the dangers of the legendary "fifth cup" of pulque—other than at specific moments in the ceremonial year. Oddly, though, those Aztecs with the misfortune of being born on the day known as "2 Rabbit" were expected to live their entire lives in a constant state of stumbling inebriation.

A few centuries earlier, traveling Europeans probably would have been surprised at the relaxed attitudes toward alcohol consumption that prevailed in the Near East, given the Koranic injunctions that inveighed against it. From the eighth through twelfth centuries, under the influence of the Abbasid Caliphate, based in Baghdad, a golden age occurred in Islamic science, medicine, and philosophy, even as Europe lingered in the Dark Ages. And the Islamic literature of the time bears witness to some pretty impressive freethinking about alcohol as much as about anything else. For long before the Romans, who were famous oenophiles, absorbed the now-Islamic lands of the eastern Mediterranean into their empire, wine drinking had been a tradition in the region. In the seventh century, the prophet Muhammad himself propounded a view of Paradise in which the rivers flowed with the precious fluid.

Although Muhammad eventually concluded that humans could not be trusted with alcohol on this Earth, others soon begged to disagree. The immensely popular late eighth-century Persian poet Abu Nuwas frequently harked back to pre-Islamic tradition in his vociferous enjoyment of wine, pointing out with immense practicality that, although wine is forbidden, God is in the business of forgiving sins. Three centuries later, his fellow Persian Omar Khayyam was still able to pen some of the most eloquent verses ever written in praise of wine. It was not until the mid-thirteenth century that a Mongol invasion and the sacking of Baghdad put an end to the era of such ruminations as "I often wonder what the vintner buys/Half as precious as the thing he sells," while a strict and humorless theology began to reassert itself throughout the core of the Islamic world.

Venturing even farther east in the late thirteenth century, the Venetian traveler Marco Polo found viticulture flourishing along the Central Asian Silk Road. He also reported encountering a "clear, bright and pleasant" alcoholic beverage in Kublai Khan's China. This was most likely a sake-like rice "beer," but some historians believe it might have been a grain-based precursor of today's distilled baijiu, particularly in light of Polo's remark that it "makes one drunk sooner than any other wine." Either way, archaeologists suggest that the distillation of alcohol was being practiced in China well before Polo's visit, though baijiu in its present form appeared only a century or two later, during the subsequent Ming Dynasty (Chapter 17).

The first sixteenth-century Portuguese visitors to the Land of the Rising Sun discovered a Japanese society that was both sake-loving and very formal, with highly stylized drinking customs especially among the upper classes. For example, while sake (ultimately derived from China) was a central feature of any fashionable Japanese social gathering, its drinking was governed by a litany of rules. The consumption of alcohol would start painfully slowly at any formal reception, as both the host and the guest of honor feigned reluctance to drink from the shared sake cup by passing it back and forth. By the end of the affair, however, all in attendance were expected to be roaringly drunk. Indeed, it was severely frowned upon if you weren't. To the astonishment of the Portuguese, business in medieval Japan was also invariably transacted over copious quantities of sake. Apparently, this custom had an entirely practical rationale: alcohol loosened inhibitions, letting the truth out and lessening the probability of duplicity.

For all the ubiquity—and popularity—of alcoholic drinks around the world, during the Middle Ages no place was more drenched in ethanol as an integral part of daily life than Europe itself. In the late eleventh century, the Islamic defenders of Jerusalem were appalled by the Crusaders' love of alcohol and the bad behaviors that went along with it, even as the Crusaders themselves measured their moral superiority by their predilection for the Eucharist wine that represented the blood of Christ.

One of the Romans' most enduring contributions to the development of European society had been the introduction of winemaking everywhere the vine would grow. Following their departure, the production and local consumption of wine continued to be popular in the countries of southern Europe. Indeed, by late medieval times the Catholic Church itself had actively developed significant commercial interests in vine growing and wine production. In the more northerly areas where wine had to be imported and was thus expensive, wine became more or less the emblem of aristocracy, as the affluent sought to differentiate themselves from the beer-swilling masses.

Following the eventual Roman retreat, northern European proletariats generally reverted to their ancient traditions of grog and beer consumption, drinking beer made from locally grown barley. This division between southern and northern drinking habits was exacerbated in the mid-fourteenth century by the onset of the Little Ice Age. By lowering average temperatures and prolonging the winters for the next five hundred years, this climatic episode made viticulture impossible in the northern regions of Europe, even as those areas remained hospitable to barley and the constituents of the meads, ciders, and various other fruit wines that continued to be produced and consumed in significant amounts.

Whatever your preferred tipple, slaking your thirst with an alcoholic beverage that had been sterilized by boiling or fermentation was likely preferable to consuming the available water, which was often of unreliable quality. With the physical decay of the old Roman system of aqueducts and drainage, the drinkability of the city water that was now mainly obtained from wells and streams had plunged. Water in the countryside, too, where domestic animals jostled for space with equally unsanitary people, was often not potable.

Whichever the preferred alcoholic beverage, the quantities consumed were prodigious. In early medieval times, when the feudal system still ruled, an average English peasant's beer allowance from the local manor or monastery seems to have been around a gallon a day—although the drink in question would in all probability have been "small beer," a weak version usually made from a second or even third pass of the mash. Predictably enough, royalty and the aristocracy were much more unstintingly generous to themselves. When England's King Edward II married in 1308, he ordered a thousand tons of claret from Bordeaux to fuel his nuptial celebrations. This was the equivalent of well over a million bottles, at a time when the entire population of Edward's kingdom had probably not long surpassed three million, and his capital city of London boasted eighty thousand inhabitants at most.

All this meant that when the distillation of beer and wine finally came along, spirits production and consumption were almost invariably imposed on existing vinous, brewing, and drinking cultures. Spirits did not replace local preferences; they simply complemented them, as wine and beer continued to be integral parts of local diets. One notable exception was in North America, where, to their surprise, the early English colonists not only encountered relatively pure water, but also resident native Americans who for the most part had no experience of alcoholic beverages of any kind. Because they had actively feared that the local water would be as bad as it was back home, the Pilgrims who landed in 1620 near Plymouth Rock on the *Mayflower* (incidentally, an old wine boat that had previously hauled claret to England from Bordeaux) had brought large quantities of beer, wine, and "strong water" to sustain them in their new environment. Once they arrived, human nature being what it is, the Pilgrims soon started feuding with the crew over who would actually drink what remained, for the sailors were naturally reluctant to set sail on a long and dangerous journey home with no ale or spirits.

Whatever kind of spirit it may have been, the Pilgrims' strong water helped them to establish relationships with the local Wampanoag residents, and they soon got busy developing their own means of producing this important resource. Dutch records from their colony of New Amsterdam, dating from 1640, already mention the making of rye whisky, while rum was being produced on Staten Island from imported Caribbean molasses as early as 1664, the year in which the English took over what would become New York. By this point, spirits had become a major trade item as well. This proved unfortunate, because the previously teetotaling locals turned out to be poor at tolerating alcohol and, instead of viewing spirits and other alcoholic beverages largely as an extension of their diet as the Europeans did, took to drinking them purely for their inebriating effects. And inevitably, among all the intoxicating beverages available, they gravitated to the spirits that got them drunk most quickly and efficiently. As a result, by the late seventeenth century colonial authorities were actively discouraging the sale of alcohol to native Americans. Eventually the practice was banned outright, although the invaders often honored such injunctions in the breach. Rum, for example, soon became the essential British trade item for beaver pelts, while brandy played that role in the more northerly regions where French influence dominated.

In the more southerly Mesoamerican region, in contrast, alcoholic beverages had long been familiar; and while the introduction of distillation techniques by the Spanish led to the local production of mezcal from an agave base, it did not cause an outbreak of alcoholism. This may have been in part because the Spanish strictly controlled the production and distribution of alcohol in their new territories, so that mezcal, which had been prized from the beginning for its gustatory qualities rather than for its ability to intoxicate, remained a prestigious beverage. In any event, the comparison suggests that the prior presence of weaker alcoholic beverages can ease the introduction of distilled spirits into society. As those back in Europe already knew, however, it was no guarantee that the undesirable effects of these drinks could be avoided.

Distilled spirits came to the Americas as beverages, but it was in another context entirely that they had originally arrived in Europe. From its Near Eastern origin in Abbasid alchemy, the notion of separating ethanol and water by exploiting their different boiling temperatures appears to have been brought to Europe in the twelfth century via Salerno, a remarkable medical school in southern Italy that brought together Greek, Latin, Arab, and Jewish medical traditions. Under its auspices a brief survey of the art of medicine was published in 1150 as part of the local version of a widely circulated compilation of medieval technical texts known as the *Mappae clavicula.* That survey, the author of which has been variously identified as "the master of Salernus," and more specifically "Michael Salerno," contained the first known European instructions for efficiently separating alcohol from water, producing a substance that would "flame up when set on fire." The knowledge of how to produce this substance was presumably considered highly proprietary, for the text describing it is both sketchy and partly in code. The product itself was evidently used in a variety of medical contexts during the several decades before the Salerno school began to lose influence, and the center of medical innovation in Italy moved northward, to Bologna.

It is still in a medical context that we next hear of distilled spirits, through the writings of the late-thirteenth-century French-Catalan physician (and probable alchemist) Arnald of Villanova. Arnald was the earliest European author to refer to spirits as *aqua vitae* (water of life), and to describe them as "the essence of wine." He also provided the first detailed instructions for distillation, and he appears to have been the first physician anywhere to have used alcohol as an antiseptic. Like his younger Catalan contemporary and fellow religious visionary Ramon Llull (a.k.a. Raymond Lully), who is often credited as the first to coin the term *alcohol* from the Arabic *al-kohl,* Arnald energetically championed the medical virtues of the substance he described as "a water of immortality." Llull took up the cause, proving to be a particularly eloquent and influential advocate of distilled spirits, which he called an "emanation of divinity . . . destined to revive the energies of modern decrepitude," and which he approvingly noted helped to fortify soldiers about to join in battle.

From this point, knowledge of distillation rapidly spread to Germany and thence throughout Europe. Large quantities of *Branntwein* ("burned wine," because it was distilled over fire) were made from the Rhenish wines that fifteenth-century Germans produced in abundance. The beverages were promoted as bracing tonics, and thus became widely sought after as medical pick-me-ups. The strongly intoxicating effects of burned wine could not long go unnoticed, and the new beverage predictably gained a mixed reputation until, at the turn of the sixteenth century, it was energetically endorsed by Hieronymus Brunschwig, a Strasburg surgeon who had built a considerable reputation in the treatment of the then-novel problem of gunshot wounds. In his influential *Book of the Art of Distilling,* first published in 1500, Brunschwig trumpeted spirits as a panacea for practically every pathology imaginable, from toothache to jaundice to bladder infections; and his readers proved only too eager to try them on their own maladies (Figure 3.1).

For all its evangelical enthusiasm, Brunschwig's book was not the first printed work on distilling. That honor belongs to *A Very Useful Little Book on Distillations,* published twenty-two years earlier, in Augsburg, by the Viennese physician Michael Puff von Schrick. For the first time this aptly titled work, which described how to make no fewer than eighty-two different herbal distillates, had brought spirits-making techniques to the public at large. And it was a rampant best-seller, reprinting thirty-eight times before Brunschwig's somewhat derivative new work took over the market. Indeed, so successful was von Schrick's handbook that, as early as 1496, authorities in Nuremberg banned home distilling in an attempt to mitigate the undesirable social consequences that were already becoming apparent.

Although such modern products as Jägermeister continue to remind us of spirits' historical origins as restoratives, distilled alcohol inevitably proved too good to remain simply the province of physicians and apothecaries. By 1495, monks at Scotland's Lindores Abbey were already distilling an "aqua vitae" that was evidently good enough to attract the attention of royalty, for in that year King James IV ordered up "eight boils of malt" that were destined for flavoring with local herbs and dried fruits, and maybe even spices. Chroniclers of the time recorded that this aqua vitae "sloeth age," and that its drinkers "abandoneth melancholy," though a recent revival can be recommended for its gustatory qualities alone, with a subtle overlay of herbs and a distinctive malt backbone reminiscent of a classic genever. Barrel-aged Scotch whiskies as we know them today were a later development (Chapter 12).

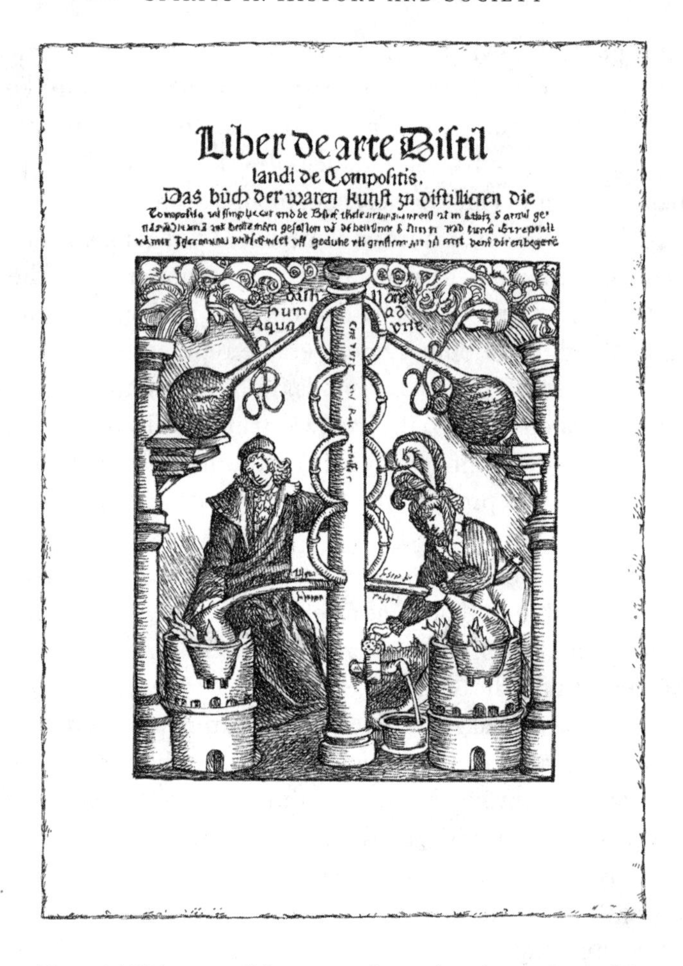

Figure 3.1. Title page of Hieronymus Brunschwig's 1500 *Book of the Art of Distilling.*

Also in 1495, the first gin/genever recipe was recorded in Holland, where the distillation of "brandewijn" was already well established (as was the production in France's Gascony region of the grape brandies that eventually became known as Armagnac). Not long afterward, the Cognac industry was founded in the nearby Charente region by Dutch merchants who were excluded from the Bordeaux wine trade by heavy taxation (see Chapter 9). These entrepreneurs needed a trade item that was more readily transportable and saleable than the thin, acidic local wines that were all the Charente traditionally had to offer, and they earned the undying gratitude of spirits lovers everywhere by urging the local growers to transform them by distillation. The era of spirits had arrived. In a remarkably short time almost every region of Europe had its own local firewater, produced from indigenous ingredients and flavorings. Most commonly the base ingredients were grains in more northerly areas, and grapes and other fruits in the south.

Over the longer term, the resulting ubiquity of readily exportable and transportable distilled spirits would have important economic and social consequences. Nonetheless, such unfortunate large-scale extravagances as England's Gin Craze still lay far in the future, and a welcome positive effect was that spirits significantly increased the variety of available alcoholic drinks, especially since they invited mixing in ways that wine and beer did not.

Because they were easily transported, grape brandies rapidly became enormously valuable commodities in Europe's expanding world trade. But no distilled beverage had greater historical, economic, and cultural consequence than rum, a spirit that, as a sugarcane derivative, could not be made in temperate Europe. During the early Age of Exploration, Europeans attempted to break the Arab monopoly on the spice trade by opening up trading routes both southward down the coast of Africa, and westward across the Atlantic. With the Portuguese in the van, such long-distance seafaring activities led during the fourteenth and fifteenth centuries to the European occupation of the eastern Atlantic archipelagoes of the Canaries, Azores, and Madeira—islands that are well suited to the cultivation of sugarcane, rum's foundational ingredient. Soon, the compact and readily transportable spirits became a valuable trade item in the coastal western African region, eventually second only to textiles.

During the Crusades, Europeans had learned from the Arabs how to cultivate sugarcane, a plant that originated in southeast Asia. But growing this tropical crop was a tough and labor-intensive business, and despite the economic rewards, the Portuguese and Spanish occupiers of the Atlantic islands were reluctant to provide the necessary labor themselves. Accordingly, as early as the mid-fifteenth century the Portuguese began to acquire slaves at their trading outposts along the African coast, shipping them west to work their island plantations. At first those acquisitions were made by raiding and kidnapping, but slaving was rapidly converted into a regular trade that penetrated ever farther into the African interior. As a result of this scaling up, by the turn of the sixteenth century Madeira's slave-fueled plantations were the world's largest source of sugar.

That preeminence would not last long. The arrival of the Genoese explorer Christopher Columbus on the Caribbean island of Hispaniola in 1492 heralded the extension of sugar plantations and the slave trade to the New World. On his second visit, Columbus took with him sugarcane cuttings from the Canaries; and within a few years the Spanish were growing cane on several Caribbean islands, and the Portuguese were cultivating it in Brazil. Because the local populations had largely or completely succumbed to European diseases, it was African slaves who worked the New World cane plantations: over the next few centuries, an appalling eleven million of them miserably crossed the Atlantic, with many more perishing en route.

Traditionally, west Africans had drunk alcoholic palm toddies and various kinds of beer made mainly from sorghum and millet, both in ceremonies and for fun. One result of this familiarity was that, from the beginning, African coastal traders were eager to obtain and quaff the strong Portuguese fortified wines and French brandies that quickly became an essential part of easing business transactions. Indeed, today's common west African custom of dashing ("tipping" or, less politely, "bribery") derives from the term *dashee,* the gift expected by a local slave trader before he even began negotiations with a European purchaser.

"Cane brandy" acquired a special place in slave trade dynamics when someone—probably a Barbados slave—figured out that this spirit could be made not just from the cane juice, which could also be converted into valuable sugar, but from fermented molasses, an otherwise mostly unwanted byproduct of the refining process. At least at first, this molasses rum must have been hugely lacking in gustatory quality: in Barbados it came to be known as "kill-devil," with early versions described by one observer as "hot, hellish and terrible liquor." But it rapidly improved enough to compete with brandy as a trade item in west Africa. Indeed, from the seventeenth century onward, rum supplanted brandy as a major barter item for the purchase of African slaves. The labor of existing slaves thus paid for the purchase of more slaves, in the "triangle trade" that involved the transfer of slaves from Africa to the New World; cotton, tobacco, and rum from the New World to Europe; and rum and manufactured goods back to Africa. In the future United States, distilleries using imported molasses flourished.

Rum also played a significant role in naval history as Britain, with its significant interests in the Caribbean, expanded its empire. During the seventeenth century, the beer rations that British naval sailors received were gradually replaced by the more compact and durable rum. For predictable reasons, this hard spirit was soon reduced to "grog" by adulteration with water and, from the end of the eighteenth century onward, with lemon or lime juice. By adding vitamin C to the Tars' diet, this supplement warded off scurvy, thus making the British, for the first time, healthier overall than rival French naval crews. (The French sailors had traditionally drunk wine containing vitamin C; but fatefully, starting in the late eighteenth century, they were instead given brandy, which was easier to transport but lacked the vitamin.) By some accounts it was the British sailors' relative freedom from scurvy that, in 1805, gave the British the edge at the Battle of Trafalgar, a clash that definitively established their naval dominance.

By the time of the Napoleonic Wars, England's Gin Craze had come and gone. But in the young United States the movement of people westward to and across the Appalachian Mountains was in full swing, and the development of the newly settled territories was facilitated by the ease with which distilled spirits could be used as a medium of exchange. Many of the new immigrants from Scotland and Ireland had brought to their new home both their whisky-drinking habits and their distilling expertise. Using native corn, which was grown virtually everywhere, they malted, fermented, and distilled readily transportable whisky. By the end of the eighteenth century, whisky had largely ousted rum as the favored spirit of the fledgling United States, becoming literally a currency of the frontier.

At the same time, whisky became a symbol of the freedom that the frontier represented. And whereas to the first colonists the notion of freedom was all about religion, from Revolutionary times onward, liberty meant freedom from taxation—the very thing that governments need most. George Washington was still president when Congress passed the Excise Act of 1791, sometimes called the Whiskey Act, which taxed distilled spirits to pay for the huge debts incurred during the Revolutionary War. This act, supported by the ale-drinking north, was vigorously opposed by both the rum-drinking south and the whisky-drinking west; and once the tax men started fanning out into western Pennsylvania and the surrounding frontier areas to collect, the insurrection that became known as the Whiskey Rebellion began. In 1794, a large military force managed to quell the insurgency, thereby establishing the federal government's sovereignty and its right to collect taxes—though, on the other side of the ledger, most observers judged that the rebels had successfully asserted the principle that government needed to be responsive to the people. And because the military action also drove some of the insurrectionists out of reach into more remote mountain regions, the Whiskey Rebellion not only helped to clarify, if not resolve, both the relationship between the federal government and the states and the federal government's responsibility to its citizens; it also ultimately gave birth to the legendary activity of moonshining (Chapter 20).

For the record, we have so far mostly spelled the word *whisky* without an "e," because this is how most people spelled it until the later nineteenth century. In 1860, however, the British Parliament passed the Spirits Act, which allowed Scottish distillers to produce cheaper blended whiskies in addition to the traditional single malts. Irish spirits-makers complained, objecting that the blends that now competed with their own product did not deserve to be called *whisky*. So, when a government commission decided in the Scots' favor, many Irish distillers started spelling their spirit *whiskey* to differentiate it in a fierce market. The upshot today is a state of benign confusion, in which the majority of distillers of grain spirits from Scotland, Canada, Japan, and other countries use the traditional spelling, while most (but not all) American and Irish distillers label their product *whiskey*. Fortunately, what you read on the label does not indicate the quality of what's in the bottle.

Just as whisky served as a currency on the western frontier in the late eighteenth century, it did much the same farther west a half-century later, as the western seaboard of the United States began to be developed. In late 1848, after gold was discovered near Sacramento, fortune hunters of every kind flooded into the new American territory of California, determined either to find the yellow metal or to fleece those who did. They brought with them copious amounts of whisky, which had great appeal in the primitive backwoods camps where the miners congregated after toiling long hours, frequently for little or no profit. As a result, people there lived in a near-constant state of inebriation. Back in San Francisco, the gateway to the Gold Rush, the key goal of separating the miners from their hard-won earnings as quickly and efficiently as possible was achieved by providing them with great quantities of whisky in a profusion of bars, bordellos, and similar institutions. As a result, the streets of San Francisco in the 1850s were, according to one press report, "hot-beds of drunkenness, and scenes of unnumbered crimes." Under Spanish and Mexican rule, California had been a relatively sleepy and pious place. Anarchic American California, in contrast, was founded on booze as much as it was on gold—and it showed.

Few of those who participated in the boom years of the Gold Rush had come to California overland from the United States. But in the ensuing decade the trickle westward from the Appalachians and Pennsylvania became a flood, as the Golden State became a terminus for the westward movement of migrants. Settlers typically started their grueling journeys with ample supplies of easily carried whisky, but the road was long and hard, and they often needed to resupply. Distilling industries sprang up in Missouri and the southwestern territories to meet the demand, providing ample quantities of booze not just for alleviating the hardships of the trail, but also for placating or trading with the Native Americans whose homelands were being traversed or invaded. Inevitably, it was the Indians who lost the most in this illicit exchange. It was still technically illegal for the settlers to supply Native Americans with alcohol, but the soldiers who were supposed to protect Indian interests turned a blind eye while one native society after another was callously pushed down the road to ruin. Meanwhile, as the nineteenth century passed its midpoint, southern slaveholders were still carousing in their mansions, plying their slaves with cheap alcohol in order to, as Frederick Douglass put it, "disgust [them] with freedom, by allowing [them] to see only the abuse of it." New Orleans, too, was energetically exporting its own brand of spirits-soaked multicultural hedonism up the Mississippi; and when the Civil War eventually came, both armies would rely heavily on spirits to maintain morale.

All in all, during the mid-nineteenth century the United States was a strong contender for the title of most alcohol-soaked society on the planet, rivaling Russia. A backlash was inevitable. Local temperance associations sprang up in cities around the nation and began to gather steam. Often allied to evangelical churches, these associations—some of which promoted moderation, while others were after all-out teetotalism—occasionally achieved their aims. But it was not until the Women's Crusade of 1873–1874 that a truly organized and nationwide opposition to the sale of alcohol got under way, fueled partly by female resentment over the rise of the saloon. Those not infrequently insalubrious establishments often excluded women, while their male patrons were seen as drinking the earnings that should have supported their families. The crusade was masterminded by the Woman's Christian Temperance Union (WCTU; still nominally in business), which rapidly made itself a political presence and popularized the immortal slogan, "Lips that touch liquor shall never touch mine!"

One major initiative involved introducing anti-alcohol propaganda into

Figure 3.2. A contemporary cartoon of Carry Nation.

school curricula. The information was disguised as science, much as creationism is in some American classrooms today. There are, accordingly, some truly sobering lessons to be learned here, because eventually local chapters succeeded in infiltrating schools in every state and territory except Arizona. Perhaps the most notable member of the WCTU was the redoubtable activist Carry Nation who, around the turn of the twentieth century, abandoned the tradition of opposing saloons by holding public prayer sessions; instead she smashed them up with an axe (Figure 3.2).

In 1893 the Anti-Saloon League (ASL) was founded in Oberlin, Ohio, but it quickly abandoned its local focus and turned its attention to lobbying Congress in Washington. Along the way it built up a powerful temperance coalition out of such unlikely bedfellows as the Ku Klux Klan, female suffragists, the Industrial Workers of the World, and the oil titan John D. Rockefeller. Ironically, the cause was significantly helped by the success of German brewers who had largely taken over the American beer industry during the second half of the nineteenth century, and who had made huge inroads into the whiskey trade in the process. The Germans were a relatively recent immigrant group against whom resentment could easily be whipped up as the Kaiser's War began. Once America was in that war, distracted legislators became easy prey to the ASL and its allies, and by the end of 1917 the Eighteenth Amendment, which prohibited the production and sale of "intoxicating liquors" in the United States, had passed both houses of Congress, overriding a presidential veto before being ratified by the states and going into effect early in 1920.

The only immutable law of human experience is that of unintended consequences. And the predictable if entirely unintended consequence of the passage of the Eighteenth Amendment was the Roaring Twenties, an astonishing period during which the new law not only failed to encourage decorum, but also made gangsterism hugely profitable, all while promoting widescale flouting of the law by ordinary citizens and police alike. The situation rapidly became so dire that it is a testament to the deep puritanical streak in the American character that it took a full thirteen years before the Eighteenth Amendment was repealed in late 1933. But those thirteen years made a deep impression on the drinking habits of Americans. During that time, for example, bulky beer gave up its prewar gains in America's drinking affections. When American beers became legal again the supply chains of malt and hops had largely dried up, resulting in a supremely insipid product. Since Americans were no longer discriminating beer consumers, the beverage remained that way until 1978, when home brewing was legalized and the craft beer movement was born. In terms of quality, if not quantity, American wine recovered a bit more quickly than beer did; but wine, unlike beer, remained a niche beverage that was consumed mainly at the upper end of the market. And it was spirits that filled the gap.

Although high-end Cognacs and whiskeys were always illicitly available to the better-heeled, the average quality of spirits sold in America during the Prohibition years was dubious at best (think "bathtub gin"). As a result, mixed drinks were popular items in speakeasies, only deepening Americans' longstanding affection for this form of drinking (for the history of mixed drinks, see Chapter 22). Affluent Americans traveled to Europe as much for the drink as for the culture, becoming well known across the Atlantic for their love of cocktails, a predilection to which their hosts enthusiastically responded.

Prohibition also had the important effect of bringing women back to the social drinking scene. Women had tended to shun the typically pretty shabby prewar saloons, but the speakeasies rapidly filled with flappers in their finest. What's more, bar owners discovered that scantily clad female singers boosted business, and it was under Prohibition in the 1920s that the bar-singing scene was born, bringing music with it. Little wonder that many aficionados look back on the Prohibition period as the greatest era of jazz.

Longed-for as it might have been, the post-Prohibition period seems to have been something of an anticlimax for Americans. This was largely due to the social effects of the Great Depression. But another unwelcome—and lasting—legacy of the Twenty-first Amendment was a chaotic system of regulating interstate trade in alcoholic products, with each state making its own rules. This led to a system for distributing alcoholic beverages that was multilayered, unwieldy, and expensive for both producers and consumers. Only the middlemen, many of them former rum-runners, benefited. Another unfortunate souvenir of Prohibition was the unusually high federal minimum drinking age of twenty-one that was conspicuously ignored by the military authorities during World War II, but that was eventually, in 1984, imposed on all the states.

One tradition that persisted despite all obstacles was the cocktail, and nightclubs with well-stocked bars dominated the public drinking scene in the late 1930s and early 1940s, at least until the outbreak of World War II, during which the army continued to march on its alcohol rations, if not quite as riotously as in the past. From the 1960s through the 1980s the art of cocktail-making waned, as premade mixes came to the fore. But eventually, improving economic conditions on both sides of the Atlantic sent drinkers back to explore the basic ingredients from which those cocktails were made. This spurred not only a revolution in creative mixology, but also, toward the end of the twentieth century, a transformation in how the spirits themselves were viewed, as more affluent drinkers and influencers of various kinds rediscovered the virtues of single malt Scotches, XO Cognacs, single-cask bourbons, and barrel-aged grappas. Despite a nostalgic interest in the classic cocktails, that trend has accelerated since the turn of this century. The happy result is that the selection of beverages available in a good bar today has never been more varied and interesting. We live in a spirituous Nirvana.

From Ingredients to Effect

4

The Ingredients

If it grows in the soil, it's likely that someone, somewhere, is distilling it. In this case, we were holding glasses of a vodka distilled mainly from spelt. This relatively unfamiliar cereal is sometimes sparingly used by brewers, but rarely by distillers, and it is famous among bakers for the nutty flavors it imparts to breads. Produced in Poland in small batches, the spring-water-based vodka gracing our table had been distilled six times and filtered through charcoal twice, making us wonder how much nuttiness would be detectable in the final product. And while we did indeed discover that nutty character, oddly enough we found it only in the creamy, compelling nose. Otherwise this was the clear, classic, smooth, and neutral spirit we had been expecting, despite an unanticipated hint of wet hay in the finish.

Ethanol, the molecule that gives spirits their kick, is found naturally around the world, but only here and there, and in relatively tiny amounts. Every once in a while a pool of liquid will be spontaneously produced that has alcohol in it, but ethanol is relatively rare because it is pretty hard to synthesize, with only a handful of natural processes able to accomplish the feat. Indeed, almost all of the ethanol on Earth results from a collaboration between plants and the tiny microorganisms known as yeasts. Plants contribute the sugars that are the chemical source of the atoms in ethanol molecules, and yeasts contribute the enzymes that convert those plant sugars into ethanol. But that's just what happens on this planet. There is also plenty of alcohol out there in the universe, in the entire absence of yeasts and plants, if you are willing to be flexible with the meaning of "plenty."

No, we haven't (yet) made contact with the Bar at the End of the Universe. But it turns out that astronomers have known for some time that there is alcohol floating around in the vastness of space. In the late 1970s, C. A. Gottlieb and colleagues reported discovering patches of methyl alcohol out there, in the following mellifluous terms: "We observed CH_3OH by its $J = 12 \rightarrow 1$ rotational transitions at 96.7 GHz toward 14 galactic sources. CH_3OH emission at 834 MHz in the galactic-center region is extended with respect to the ~40′ telescope beamwidth although the column density of CH_3OH molecules peaks near Sgr A and Sgr B2." Roughly translated, this means that the Gottlieb group's radio telescope found methanol (CH_3OH, a close structural relative of the ethanol molecule) in fourteen separate distant galactic centers. And while drinking poisonous methyl alcohol is not recommended, its detection in the space between larger blobs of matter out there in the universe was a significant scientific discovery. In 1995, S. B. Charnley and colleagues wrote a paper entitled "Interstellar Alcohols" in which they described an abundance of alcohol concentrations stretching across the known universe. A few years later, a cloud of methanol was detected that measured 288 billion miles across, shrouding what is called a stellar nursery. The important point here is not that nascent stars are alcoholic. Instead, the discovery added to our understanding of how stars form, since the methanol appears to have played an important role in the process whereby massive stars spin off stellar nurseries.

It also sparked a series of discoveries that make it clear the known universe is full of alcohol. One remarkable kind of interstellar alcohol is known as vinyl alcohol (C_2H_4O). It forms a lot of isomers (molecules that have the same formula—C_2H_4O—but that assume different shapes because of the way the atoms are attached to each other). There are 125 kinds of small molecules floating around in space, and most of these are made of six or fewer atoms. These molecules form when small molecules, and sometimes just atoms, collide—the smaller the molecule, the more easily it is formed. Formaldehyde (H_2CO), for example, has only four atoms, so it fits in this category nicely. Methanol has six, so it is also pretty easy to form. But vinyl alcohol is over the six-atom limit, and hence is difficult to form through simple gas-phase chemistry. Ethanol, too, contains a whopping nine atoms (C_2H_5OH). Larger molecules like these need an extra push to make them cohere. Understanding what gives this push is important if we want to explain how more complex molecules can form.

Researchers think the extra nudge needed comes from interstellar dust that smaller molecules can stick onto during synthesis. That is the good news. The bad news is that, with one notable exception, the alcohol molecules are so sparsely distributed in space that to collect a shot glass full of pure ethanol one would have to hold that glass out of the cargo bay of a spaceship for well over a half-million light years: a distance equivalent to the entire breadth of our own galaxy, the Milky Way. The exception? A comet known as Lovejoy, which Nicolas Biver and his colleagues say releases "as much alcohol as in at least 500 bottles of wine every second during its peak activity." Still, comet chasing is not a very efficient way to obtain a shot of booze, and fortunately for Earthbound humans who enjoy spirits, conditions on our planet allow for other ways of obtaining ethanol.

The starting ingredients of any spirit are pretty basic. Water, most of which the distiller wants to get rid of eventually, is the simplest. Water is used in three phases of spirits production: in making the mash from which the spirit is distilled; in the redistillation process (where it is used to dilute neutral grain alcohol); and finally, in diluting or "proofing down" the finished product. Fortunately,

LOCATION	DESCRIPTION	pH
		SOFTER
FRANCE	MINERAL SPRING WATER	5.5
ENGLAND	DEMINERALIZED DOMESTIC WATER	7.0
GERMANY	SPRING MINERAL WATER	7.1
ENGLAND	RECYCLED DOMESTIC WATER	7.2
FIJI	ARTISANAL SPRING VOLCANIC MINERAL WATER	7.7
ICELAND	VOLCANIC GLACIAL MINERAL WATER	8.4
		HARDER

Figure 4.1. Water used in the London Craft Distilling Expo experiment. Water from all over the globe was used to distill gin, revealing surprising differences in overall taste that correlated with pH and water hardness or softness.

water is everywhere on this planet. It is so simple and ubiquitous, indeed, that we often take it for granted. But not all water is the same, because nowhere is the water entirely pure. Wherever you might collect it, other things invariably come along with it, either in solution or in suspension. This extra stuff influences what is called the "softness" or "hardness" of the water. Harder water has a lot of minerals mixed in with it, while softer has fewer. This makes a difference to the distiller, because the concentration of the minerals in water affects the potential for the hydrogen in it to react with other atoms. This potential (some call it the "power") of hydrogen is abbreviated as pH. Low pH indicates low potential for hydrogen activity, showing that the solution is acidic. In contrast, solutions with a high pH are "basic," and have high potential for hydrogen activity. In general, water hardness and pH are correlated, because harder water will contain more minerals that act as a buffer and raise the pH.

Attendees of the 2015 London Craft Distilling Expo were treated to a talk about an interesting experiment that tested the impact of the source of water on gin production. Six batches of gin were produced, using the different waters shown in Figure 4.1. The experimenters used the same procedures, the same botanical ingredients, and the same distilling equipment (a pot still) to make the gin.

The different gins produced had discernibly distinct characters and tastes when evaluated by a panel of humans—whose reactions are notoriously subjective. Nonetheless, it was agreed that the gin made from French mineral water at pH 5.5 was "soft and clean, with some bright, floral notes coming through, as well as clean and crisp angelica and pine . . . a nice flow of flavors with a long, lingering, dry finish. Very sippable, with a silky and light texture." In contrast, the gin made from the Icelandic water at pH 8.4 was "fruity on the nose, but . . . has a very short flavor profile, which is over almost immediately. There is a little burst of juniper that quickly disappears. It also has a cloying texture, is rather hot at the end, and overall is not very pleasant to drink." The taste and character of the gins made with water in the pH range of 7.0 to 7.7 differed too, but not as radically as the gins made with the Icelandic and French water, which were at the opposite ends of the pH scale. Doubtless the waters used also differed in features other than pH, and the differences between gins made with the various waters mostly involved texture and mouthfeel, not taste. But despite the distillers' initial conclusion that water had only a marginal influence on the quality of the gin-drinking experience, it is tantalizing to wonder about water's effects—especially because when the six gins were all allowed to age, the differences became even more distinct. It seems that water choice may indeed be rather significant when making spirits, and that makes sense. After all, the water used in the proofing stage alone will make up around 50 percent of the finished product.

Next in the list of ingredients comes sugar, the source of atoms for the small alcohol molecules that the distiller wants to end up with. Sugar for distilling comes from many sources, virtually all of them plants, mainly cereals (edible grasses) or plants that bear fruit. Plants have strange lives compared to us animals, starting with the fact that to get energy for their survival, they use sunlight in a process called photosynthesis. This process can be summarized in the following equations:

$$6CO_2 + 12H_2O + \text{Light Energy} \rightarrow C_6H_{12}O_6 + 6O_2 + 6H_2O$$

or

$$\text{carbon dioxide} + \text{water} + \text{sunlight} \rightarrow \text{sugar} + \text{oxygen} + \text{less water}$$

Plant cells synthesize sugar in tiny organelles called chloroplasts that are found mainly in the cells of the leaves and give plants their green color. The resulting sugar is then transported to the rest of the plant via the vascular system, and used

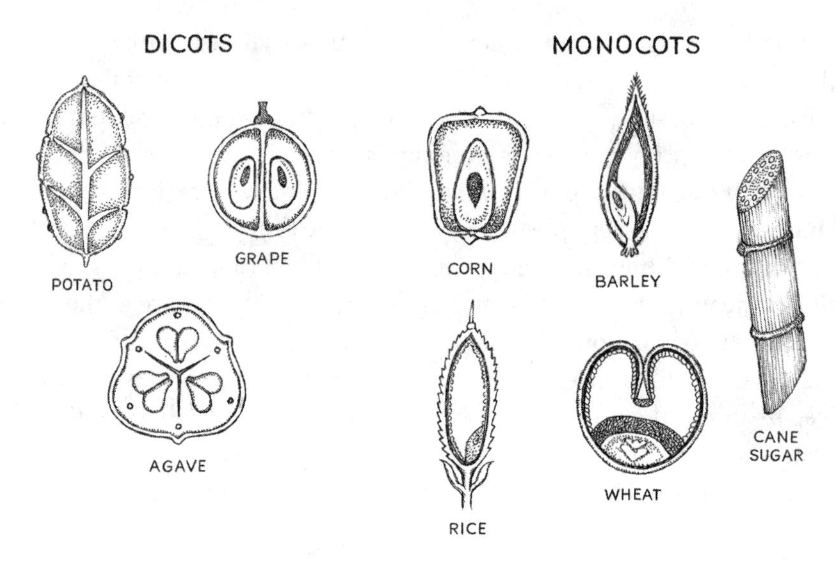

Figure 4.2. Plant sources of sugar. The three diagrams on the left are of dicots and the five on the right are of monocots.

as a nutrient for the plant as it grows and reproduces. Many plants also surround their embryos with sugary substances.

Plants with flowers can be separated into two major groups, known as monocots and dicots. From their names it should be obvious that monocots have one of something, whereas dicots have two. Those things are cotyledons, little leaflike structures that bud off of the developing embryo and serve as "nursemaids" that provide a source of nutrients for the developing plant. Most plants are good parents, sequestering a lot of sugar around or near their seeds (Figure 4.2). Grapes famously have a sugar-rich fleshy layer surrounding their embryos. Other dicots used by distillers, such as agave, also have large amounts of sugar surrounding the developing embryo. Monocots like the "basic grains"—corn, barley, wheat, rice, and so forth—also have sugary layers around their embryos. Sugarcane, on the other hand, stores sugar throughout its stem structures.

But nourishment of the developing embryo is not the only reason that plants produce sugars. Some plants also produce sugar to attract seed dispersers. The sweeter the fruit, the more likely it is that an animal like a bird will ingest it, fly away, and poop out the seeds to germinate somewhere else. Everyone benefits.

A grape is an easily available reservoir of sugar—squeeze it and you will have all the sugar you want. In contrast, most grains have to be tricked into giving up their sweetness. Barley, for example, has evolved a system in which the starches in the grain are released only when germination occurs and the developing embryo needs to consume the sugars derived from them. Consequently, brewers have had to figure out how to release the starch before too much of it has been used up by the plant. It's a double whammy of a trick that has been perfected over the long history of brewing. First, the malt-makers (maltsters) soak the grain in water, which tricks it into germinating prematurely. Then they quickly dry the grain with hot air to stop the embryo from growing, thus leaving most of the starch available for the brewer. Just as important, the grain malting process releases enzymes called amylases. These break down the starches in the grain into the sugars—such as glucose—that are the key molecules for the fermentation process to come. Barley is particularly good at doing this, which is why some of it is often included in mashes for spirits that otherwise mostly use other grains. Significantly, too, whatever the grain, we don't usually know where it was grown. In this sense, spirits are unlike wines, which come from grapes whose origins are carefully tracked.

Ethanol is produced in nature when any sugary concoction interacts with the single-celled organisms we call yeast. Yeasts are fungi, closely related to mushrooms—and they are thus, oddly, more closely related to us animals than they are to plants. Many yeast species have evolved a way to process sugars for energy and convert them into alcohol. The specific process is described in this chemical equation:

$$C_6H_{12}O_6 \rightarrow 2C_2H_5OH + 2CO_2$$

or, more simply:

glucose → 2 ethanols + 2 carbon dioxides

These reactions occur inside the single yeast cell, using specialized enzymes. The sugar (glucose) is essentially the yeast's food. After converting the sugar into energy, the yeast transports the resulting alcohol beyond its bounding membrane (to deter competitors) and gets on with life. As we explained earlier, ethanol is a toxin; in high enough concentrations it is toxic even to the yeasts themselves.

This limit caps the strength of naturally occurring ethanol on our planet at about 15 percent by volume. There are tough yeasts that can tolerate up to 25 percent ethanol by volume, but such "yeasty Hulks" are both rare and the result of human intervention, having been bred by brewers specifically to increase the amount of ethanol they can squeeze into their concoctions.

Brewers have tended to be a bit more proactive than winemakers in their use of yeasts, developing hundreds of yeast strains for brewing the many kinds of beers that now exist. But remember that this has been a trend only for the century and a half since Louis Pasteur discovered yeast's role in fermentation. Thus, when the Bavarian authorities legally defined beer in the *Reinheitsgebot* law passed in 1516, they specified only three ingredients: barley, water, and hops. Note the conspicuous absence of yeast; 1516 was way before Pasteur's time, and it had to be added to the definition later. Once brewers had figured out that yeasts—and one species specifically, *Saccharomyces cerevisiae*—were important in brewing, they started to match their yeasts with their brews, and in essence domesticated these single-celled organisms for their own purposes.

Over the past five years, researchers have been able to obtain the genetic blueprints or genomes of millions of organisms, and our knowledge of yeast has benefited from this surge. For winemakers and beer brewers especially, the new information will help to make better products. Further, one discovery by genetic researchers should draw the attention of creative distillers: while all wine yeasts look pretty much alike, beer yeasts are wildly different from one another. This phenomenon was actually not unexpected, and it is due to the relative degrees of domestication and wildness of wine and beer yeasts. Interbreeding through domestication, as we see with wine yeasts, makes genomes more similar, whereas allowing the cross-breeding of genomes (as happens with beer yeasts) makes them more varied. The taste effects of yeast will thus be more variable when beers are being made, and less so when wines are. That's something for distillers to keep an eye on, too, as they seek to expand their taste options for spirits.

Thanks to yeasts, alcohol is found naturally all over our planet. But from the spirit-maker's point of view, the problem with those little pools of natural ethanol is that they come with a lot of impurities that derive mostly from the fermentation process. So in making spirits, the initial fermentation of the wash is just as important as the subsequent distillation step that removes the unwanted elements.

As we've seen, the "discovery" of natural fermentation has occurred many times over, and not only by humans but also by several non-human animals. Even fruit flies seek fermenting fruit. Humans are different, however, in having put their discovery of fermentation to work by creating wine, beer, and other beverages that feature low concentrations of alcohol. We have explained wine making and beer brewing in other books, but it's worth reviewing these processes here because, for most spirits, the starting point is a liquid with a lower initial concentration of alcohol, commonly a brewed mash or a fermented wine. After all, the distillation process doesn't create alcohol, but rather concentrates it. Naturally enough the nature of the starting liquid will deeply affect the distilled end product, so although there is a huge difference between the finished products of brewers and winemakers and the starting materials for the still, a good spirits producer will probably be familiar with the ins and outs of wine production and beer brewing.

Beer brewing, in particular, can be quite frustrating for a beginner who is not familiar with sterile techniques. This is because any sugary mixture that becomes infected with bacteria will sour badly. Bacteria have evolved to break down sugar into acetic acid, contamination with which is a death knell for a beer batch. As a result, most brewers will attempt to make their product as bacteria-free as possible. In fact, if you look at the brew as an ecosystem, a modern brewer usually tries to include a monoculture yeast strain with only a single species present (although, to prove the rule by exception, lambics and sour beers of all kinds are packed with different kinds of microorganisms). The whisk(e)y distiller, in contrast, strives to include many species during fermentation, because the sourness derived from bacterial breakdown of sugar is what gives the distillates much of their character and flavor. Remember also that most beer today includes hops, which not only limit souring, but also add a rather pleasant bitterness to its taste and inhibit the growth of bacteria. Brewers nowadays also boil their brews to produce what is called wort or, more descriptively, "sugar water." There are two important reasons for boiling: to disinfect the wort, and to break down the long-chain sugars derived from the grains into the smaller glucose molecules that drive fermentation. Boiling, however, also drives off potentially desirable aromatics.

In contrast, any whisk(e)y distiller knows better than to add hops to the mash, and most distillers don't even boil it. Skipping these steps increases the microbial diversity of the initial mash, adding elements that the distiller may want to preserve in the final product. For grain-based spirits distillers, the length of the fermentation step is also critical to retaining an ecological balance within the mash.

Most distillers will pitch their own yeast on the mash, giving those microbes of choice a head start in the ecology of early fermentation. But as the alcohol level in the mash increases, the yeast tire, become sluggish, and even can die off if the alcohol concentration becomes too high. When this occurs, other microbes start to compete for the sugar that is still left in the mash. And even though those bacteria will ferment sugars into sour acetic acid, that might be just what the distiller wants. After all, most unwanted compounds can be removed by the next step, distillation, while any sour stuff that develops in the mash during fermentation might add to the flavor of the spirit.

By now you know that ethanol occurs naturally on our planet as a result of yeast fermentation, and that humans have piggybacked on this natural process in order to make drinkable spirits. But what if we could cut out the middleman, yeast, and use basic chemical reactions to make ethanol?

Ethanol is not a very complicated molecule. It has two carbons, six hydrogens, and one oxygen atom joined together. But what *can* become complicated is that, in three dimensions, the same combination of atoms may flip and flop itself into all manner of different molecules. And it is the shape of a molecule that determines its function. For instance, both ethanol and dimethyl ether have the same number of carbon, hydrogen, and oxygen atoms—two Cs, six Hs, and one O. But in structural and behavioral terms they are very different. They are technically called isomers, and, if they are playing different functional roles, they are called functional isomers. Figure 4.3 shows the three-dimensional "stick and ball" models for ethanol and dimethyl ether. These models demonstrate nicely the differences in structure of these two isomers. Dimethyl ether is pretty symmetrical, while ethanol is rather lopsided. Because molecules often act like locks and keys, the overall shape of any molecule dictates how it will react with others, and the differences here are striking. Ether is an organic chemical that is relatively insoluble in water, while in contrast ethanol is what is called "infinitely" soluble. Ether has been used by humans as an anesthetic, putting someone who inhales it to sleep quickly and efficiently (if not entirely pleasantly), and ethanol has a whole variety of uses. Oddly enough, even though the two molecules have the identical number and kinds of atoms, ethanol is classified as an alcohol while, for reasons that will soon become clear, ether is not.

But now take the ethanol molecule in Figure 4.3 and remove the carbon

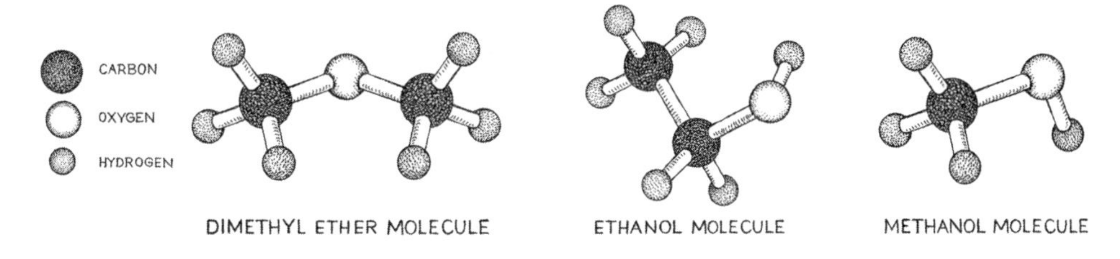

Figure 4.3. Stick and ball models of dimethyl ether, ethanol, and methanol molecules.

with three connected hydrogens from its left end. When this carbon is removed, two of the hydrogens go away too. This is a result of the atomic bookkeeping that automatically maintains chemical balance. When this "methyl group" is removed, we get CH_4OH, or methanol. Methanol looks like the truncated ethanol that it is. But its shape is so different from ethanol that the functions of the two molecules are quite different. While humans can tolerate moderate amounts of ethanol, a tiny amount of methanol will wreak havoc on the bodies of any vertebrate, including us. Oddly, even though the two molecules—ethanol and methanol—are so different functionally, they are both called alcohols. Here's why. If you look closely at their stick and ball models, you will notice that both ethanol and methanol have a hydroxyl (OH) group coming off their right ends, while ether does not. Alcohols are defined chemically as small organic molecules in which one or more OH groups are attached to a carbon atom. Table 4.1 shows this common feature: there you'll notice several alcohol molecules, all having different numbers of carbons and hydrogens, but all with a hydroxyl group (OH) on one end.

Alcohols are categorized in many ways. One of the simplest is to separate all alcohols into three "flavors": isopropyl alcohol, methyl alcohol, and ethyl alcohol. Humans can tolerate only ethyl alcohol, because the other two are toxic to all animals. Another way to categorize alcohols is by the way in which the hydroxyl group attaches to the rest of the alcohol molecule. This results in recognizing primary, secondary, and tertiary alcohols. Remember that, as the "ultimate bookkeeper," nature allows a carbon to have four other atoms attached to it. Primary alcohols are those in which, like ethanol, the hydroxyl group is attached to a carbon, which is in turn attached to only one other carbon atom (Table 4.1). Secondary alcohols are those in which the hydroxyl group is attached to a carbon that is in turn attached to two other carbon atoms, and tertiary alcohols have three carbons attached to the hydroxylated carbon (Figure 4.4).

Table 4.1. STRUCTURES AND NAMES OF SOME PRIMARY ALCOHOLS

FORMULA	NAME
CH_3OH	methanol
CH_3CH_2OH	ethanol
$CH_3(CH_2)_3OH$	butanol
$CH_3(CH_2)_3CH_2OH$	pentanol
$CH_3(CH_2)_4CH_2OH$	hexanol
$CH_3(CH_2)_6OH$	heptanol
$CH_3(CH_2)_6\ CH_2OH$	octanol
$CH_3(CH_2)_9OH$	decanol

How can we synthesize ethanol from smaller molecules? Because there are so many kinds of alcohols and related compounds, the process is neither easy nor very efficient. A lot of the artificial ethanol produced is in fact a product of petroleum, because one of the major components of petroleum is ethylene (ethene) or C_2H_4. When combined with steam (gaseous H_2O) this simple hydrocarbon will form ethanol, in an exothermic reaction that releases energy. But this reaction is not very efficient: typically only 5 percent of the ethylene is converted to ethanol. Even so, each year about two million tons of ethylene-derived ethanol are produced industrially worldwide, with most of the product used as disinfectants, solvents, or fuel.

Scientists have come up with a variety of ingenious ways to synthesize ethanol. In 2018 Qingli Qian and colleagues managed to synthesize it from its isomer,

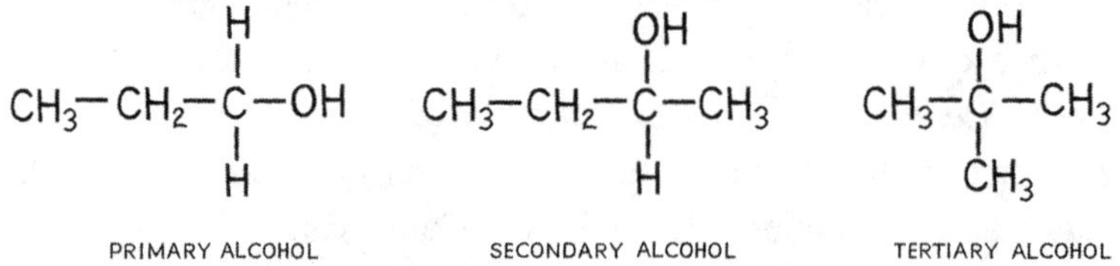

$CH_3-CH_2-C(H)(H)-OH$ PRIMARY ALCOHOL

$CH_3-CH_2-C(OH)(H)-CH_3$ SECONDARY ALCOHOL

$CH_3-C(OH)(CH_3)-CH_3$ TERTIARY ALCOHOL

Figure 4.4. Structures of primary, secondary, and tertiary alcohols.

dimethyl ether, by making it react with carbon dioxide and hydrogen. The conversion appears to be reasonably efficient, but its practical utility remains to be seen. Syngas (synthetic gas) has also been suggested as a source for alcohol. This fuel gas, which is made up of hydrogen, carbon monoxide, and some carbon dioxide, can be produced from a wide range of source materials such as coal or natural gas, and it is already important for the synthesis of ammonia, methanol, and pure hydrogen. Researchers were able to use rhodium catalysts to boost the production of ethanol from syngas, but at a somewhat disappointing rate until, in 2020, Chengtao Wang and colleagues succeeded in synthesizing ethanol directly from syngas with a different approach that used as a substrate highly reactive zeolite crystals—beautiful, pale-colored, soft crystals found in relatively recent volcanic fields. After being crushed into a powder, these crystals can be used to boost the production of artificial ethanol significantly compared to earlier methods. All of these synthetic approaches to alcohol production circumvent the time- and energy-intensive biological pathways to making ethanol, and they have thus attracted researchers interested in the production of consumable alcohol. After all, they promise to produce the ultimate "neutral spirit," consisting entirely of ethanol molecules.

The notion of synthesizing a product to replace the alcohol that has traditionally been distilled brings us to the subject of synthetic alcohol substitutes. Enter David Nutt, the "Dangerous Professor." Professor Nutt has made a living out of arguing for recreational highs. It isn't that he is an alcoholic, or a pothead, or a crack fiend. Rather, he has spent much of his career trying to understand the use of recreational substances in modern society, as a "fierce advocate of what he says are more enlightened, rational drug policies," as one observer put it. In 2009 Nutt was fired from his position as chair of the United Kingdom's Advisory Council on the Misuse of Drugs for saying something that, he points out, Barack Obama also noted: namely that marijuana is less potentially harmful than alcohol and a whole slew of other drugs. On hearing Obama's declaration, Nutt issued a memorable caution: "At last a politician telling the truth. I'll warn him though—I was sacked for saying that." His own statement about cannabis had upset Alan Johnson. Johnson, who was then Home Secretary (a position much like the U.S. Secretary of the Interior), had just upgraded marijuana from a class C to a class B drug (the greater the perceived danger of the drug, the higher the grade). Nutt knew that this didn't make sense, because *Cannabis,* as many U.S. states have recognized, is probably one of the safer drugs in use.

The Dangerous Professor also has ideas about the effects of alcohol. He correctly recognizes that alcohol, consumed in excess, can be toxic to the human body. But he also argues that if a synthetic substance could be produced that mimics the effect of alcohol without its side effects, many of the unwanted effects of alcohol would disappear. We will see how alcohol affects the brain in Chapter 8, but as a sneak peek, we point out that alcohol inebriates by targeting specific chemicals that are involved in normal brain function. Although alcohol's effects on the whole brain and the neurotransmitters are not easy to duplicate, Nutt suggests two compounds that may be particularly promising as substitute inebriants. Even better, antidotes could be administered to counter their inebriating effects, allowing one to hop in one's car and safely drive home after a night of drinking.

The spirits lover in all of us will, of course, be wary of this suggestion. Why desert the beverages you have loved for most of your life for some unknown chemical substitute? And what about the legal and ethical problems in Nutt's proposal? Not to mention that the flavor components in your favorite spirit are overwhelmingly determined by the congeners, those nonalcohol molecules that the distiller either allowed to survive the distillation process or added during the aging-in-wood step. But Nutt carries on undaunted. He adds flavor with fruit juices, and as of 2019, he had found some great candidate molecules, some of which he even tested on himself. He calls the new category of substances *alcosynths,* and his product Alcarelle might be available by 2025. Stay tuned—perhaps with a Seedlip cocktail to make the wait more enjoyable.

5

Distillation

We were surprised that anyone would make a vodka in Frankfort, Kentucky. But this product of the Buffalo Trace Distillery, famous for its bourbons, advertised that "this is how vodka is supposed to taste." Naturally, we were curious. According to its creator, the wheat-based spirit in the bottle we obtained is batch-distilled a modest seven times in an impressively idiosyncratic "one-of-a-kind micro-still," and three times more in a "vodka pot still," before being triple-filtered. We write "modest" because the same distiller once made a vodka that allegedly went through the still a mind-boggling 159 times—

although we are still not sure exactly what that meant in practice. The total of ten distillations of our sample should have been more than enough to assure pristine purity, so we were mildly surprised when we encountered nuances of rosemary and lavender on the nose. The palate and finish were as smooth as we could have hoped for, with no vodka burn at the end.

How do you go from a mixture that contains lots of stuff (literally) to a rather pure alcohol? As we have explained, most spirits start from a "mash" that is a complex mixture of all kinds of tissues, molecules, and atoms, usually derived from plant sources, as well as microbes, yeast, water, and other molecules and chemicals that come from the environment. It is truly a wonder that humans have figured out how to get from this spectacular stew to the highly alcoholic and relatively pure spirits that we drink.

To understand it, some chemistry is helpful. Living things on Earth are carbon-based, partly because they have a lot of carbon atoms. Yet carbon is just one of 118 elements found on our planet, if we include the twenty-seven elements that can be observed only after chemists and physicists have performed some magic in their laboratories. Except for the rather suspiciously named *technetium,* all of the first ninety-two elements in the periodic table created by the alcohol expert Dmitri Mendeleev are found naturally on Earth (see Chapter 10 for more on Mendeleev). But living things are selective about the atoms they use in building their bodies. Most animals incorporate only six, among which carbon is the second most common. The complete list, in order of their abundance, is oxygen (O), carbon (C), hydrogen (H), nitrogen (N), phosphorus (P), and sulfur (S). High-school students are taught the mnemonic OCHNPS to help remember this.

As you'll recall, the organisms responsible for almost all the alcohol on the planet are the yeasts, and these single-celled eukaryotes (that is, organisms with a nuclear membrane surrounding their genetic material) have evolved to use these same big six atoms plus a pinch of chlorine. Finally, plants use not only the big six, but also four other atoms: magnesium (Mg), silicon (Si), calcium (Ca), and potassium (K). That does it for the living ingredients in the mash; but our last and most abundant ingredient is water—which, of course, is a nonliving and less complex entity. It has the fewest elements of any of our spirits ingredients (just H and O); but, as we saw, water can carry a lot of different minerals in solution. As a result, there may be atoms of very many different kinds in the mash, all of which can interact to make molecules. Which means that, to understand distillation, we need to know something about atoms.

An atom has a nucleus, or central core, with two kinds of particles in it: neutrons and protons. Around this central core, small particles called electrons circle (for lack of a better word), moving in different orbits known as "orbital levels." To understand atoms, you also need to understand the mysterious quality known as "charge." Luckily there is only one kind of charge we have to worry about, namely electrical charge. Neutrons have no charge, so only the electrons and protons in an atom are relevant here. Anything that has the same number of protons as electrons is neutral with respect to charge. But if something has more electrons than protons it is negatively charged, and with more protons than electrons it is positively charged. Anything that is net positively or net negatively charged is called an ion. Any atom can also be characterized by its atomic number, which is the count of balanced electrons and protons it has. For instance, sodium (Na) has an atomic number of eleven, because in its balanced state it has eleven electrons and eleven protons. If a sodium atom loses an electron, then it is a positive sodium ion and is represented as Na+. The charge of atoms is the currency that nature uses to keep score of balance in the universe, and it is a strict bookkeeper.

The structure of an atom or an ion is dependent on how many electrons it has, and where those electrons are with respect to the various orbital levels. For instance, a sodium atom has eleven protons in its nucleus, and eleven electrons orbiting around the nucleus in three different orbital levels. Sometimes the electrons of one atom can be transferred to an adjacent one, while preserving stability. Such transfer decreases the number of electrons in the donor atom and increases the number of electrons in the acceptor atom. When this happens, the electrical charges of the two atoms change and produce a balanced state between the two. The donor becomes positive, and the acceptor becomes negative, and they are thus both said to be ionic.

As its equation suggests, water (H_2O) has two hydrogen atoms and one atom of oxygen, and it is incredibly stable. The story of water is literally as old as the universe. Hydrogen atoms are highly reactive with others and so, not surprisingly, in the early universe they found other hydrogens and formed the more

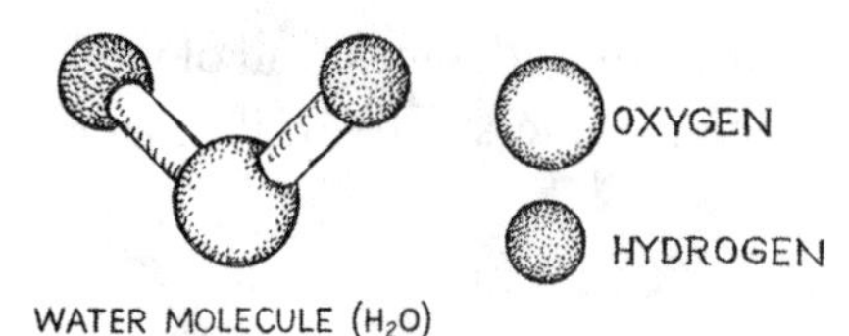

Figure 5.1. Ball and stick model for a water molecule (H_2O).

stable hydrogen (H_2) molecule. Oxygen is also reactive, and readily forms a molecule with other oxygens, making the more stable O_2. It might seem logical to simply add H_2 to O_2 to end up with water, but it isn't that simple. Energy is needed to get this reaction started, and an intermediate molecule is formed: H_2O_2, or hydrogen peroxide. This molecule is very unstable, decomposing into water and oxygen molecules, so that while the balanced equation for the formation of water from oxygen and hydrogen molecules is $2H_2 + O_2 \rightarrow H_2O$, this doesn't really demonstrate the complexity of how water was formed early in the history of our universe.

As we saw earlier, chemists like to use "stick and ball" models to describe the bonds between atoms (the sticks), and the atoms themselves (the balls, of differing size and usually color). See the stick and ball model for water (Figure 5.1). Another way to write the structure of water is to use lines to represent the bonds, and letters to represent the atoms (Figure 5.2). This also allows the chemist to say something about how the atoms are spatially related to each other, such as the angle between the two hydrogens (about 105 degrees). Both of these ways of representing a molecule, along with the standard equations used earlier, allow chemists to make sure that the balance demanded by nature exists in their molecular structures. But we also need to know something about the different atoms. For instance, hydrogen (H) must have one bond coming off it, while oxygen (O) must have two bonds, and carbon, the interesting atom of organic things, must have four bonds coming from it. You see this clearly in both figures. We will discuss the structure of sugar and alcohol in detail later, but here is a sneak peek at ethanol,

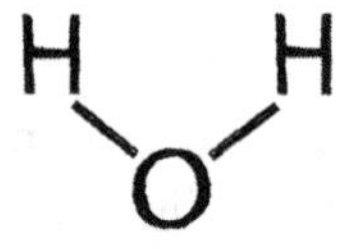

Figure 5.2. Diagram of a water molecule (H_2O), using lines and letters.

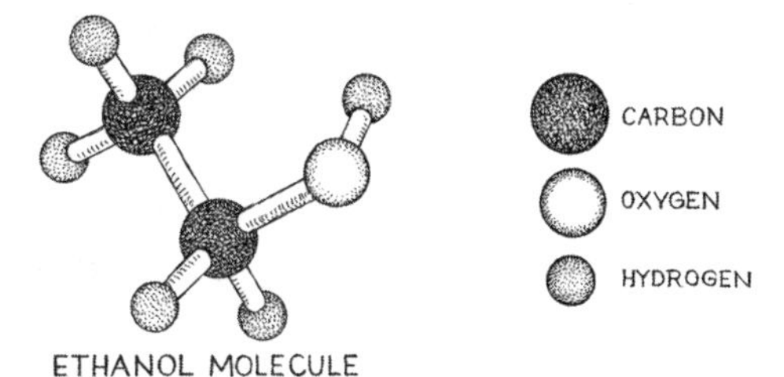

Figure 5.3. Stick and ball model of an ethanol molecule (C_2H_5OH).

the drinkable kind of alcohol, with black representing carbon, white representing oxygen, and gray representing hydrogen (Figure 5.3).

Note that the two carbon atoms have four bonds each coming from them; one is between the two carbons, another is between a carbon and an oxygen, and the rest are with hydrogens. Bonds made between the atoms in molecules may be of three basic types: ionic, covalent, and metallic. Metallic bonds occur between metals, and we won't be concerned with them here. Ionic bonds are the strongest and most difficult to break, because they are formed by the donation of an electron by one atom to another. Sodium chloride (NaCl), or table salt, is a case in point: it is a compound in which the chlorine atom donates an electron to sodium, creating the "lattice crystal" illustrated in Figure 5.4. Weaker covalent bonds occur where the interacting atoms actually share electrons. All of the bonds in the ethanol molecule are covalent.

The ethanol molecule is pretty stable, meaning that it survives in solution

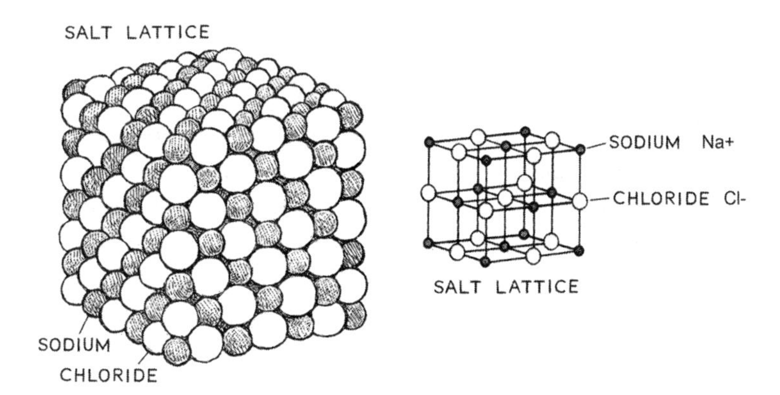

Figure 5.4. An NaCl lattice formed by ionic bonds between sodium and chloride.

with other atoms and molecules over a relatively broad range of conditions. The good news is that it has this solubility; the bad news is that this solubility makes it difficult to purify. As mentioned earlier, ethanol is infinitely soluble. In other words, it is miscible, meaning that it mixes with other solutions in any proportion. Thus, when ethanol and water are put together, the two solutions mix to create a single solution. By contrast, if you mix water with decanol (another alcohol, with ten carbon atoms), the decanol will separate from, and float atop, the much denser water.

Molecules can coexist in one of three states, or phases: liquid, gas, or solid. Both temperature and pressure are important for determining in which phase a compound will be at any time. Mostly when we talk about these three phases, we are assuming the pressure is one atmosphere (atm), defined as the pressure at sea level on Earth. Because water is so central to the distillation process, we will spend some time discussing its three states. At room temperature and one atm, water is a liquid. At one atm and zero degrees Celsius (or C), liquid water transits to the solid phase and forms ice. At one atm and 100 degrees C, liquid water transits to the gas phase. On the top of Mount Everest, water will transit to the gas phase at 71 degrees C, because the pressure is lower there (0.3 atm) and the boiling point goes down as atmospheric pressure decreases. In contrast graphite, or solid carbon, is always a solid on the Earth. It transits to the gas phase at 3,550 degrees C, and there is no place on the surface of the planet where the temperature and pressure are right for it to change phase.

Any solid consists of packed molecules, while gases are made up of diffuse distributions of molecules. This structural arrangement means that solids are denser than liquids, which in turn are denser than gases. Dense molecules sink in solutions, whereas less dense molecules float. This means that a solid placed in a liquid made of the same molecule normally sinks, and a liquid mixed with the gaseous form of a molecule will settle below the gas. Think of that lattice formation after the ionic bonding of sodium and chloride to make salt. The tightly packed atoms in that lattice structure make it much denser than water, so that when you add a little salt to a glass of water, it sinks to the bottom. Eventually it will dissolve, but that's another story. Figure 5.5 shows schematically how this works with oxygen.

But wait a minute! That cocktail you are drinking has ice floating on it, and

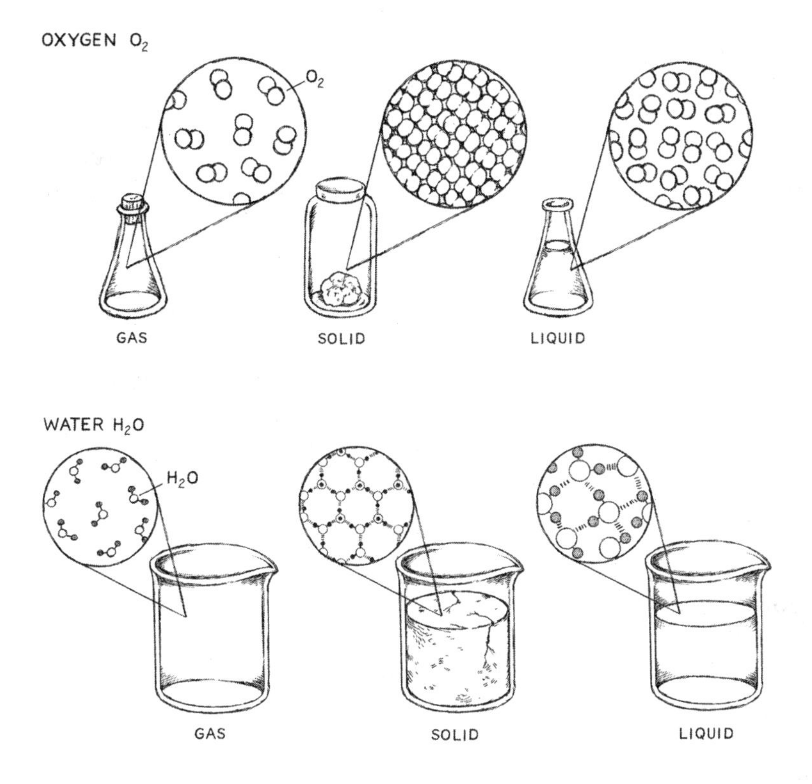

Figure 5.5. Oxygen molecules (O_2) in, from left to right, gas, solid, and liquid states.

ice will also float just as well in a glass of pure water at room temperature. Why does ice, a solid, float above liquid water? Because as the temperature drops, the bonds made between the hydrogen and oxygen atoms stabilize the interactions of water molecules. As the temperature is lowered the water molecules stop moving, so they can't form complex lattices or pack together like sardines. The hydrogen bonds in water don't allow the cold molecules to get as close to each other as they are at room temperature. This reduces the density of solid water to only about 90 percent of that of liquid water, which is enough to push the ice to the top of your cocktail.

That ice floats is incredibly important for life on Earth. Lakes freeze over, with the ice on top insulating the water below. If that ice were to sink, entire water bodies would freeze from the bottom up, making it impossible for most of the organisms living in them to survive; and complete thawing would be unlikely when temperatures warmed up. In addition, surface ice reflects a lot of sunlight, a process that is important for regulating global temperatures. It does this particularly efficiently at the poles, which is an excellent thing because if ice sank in the oceans, there would be no ice surface at the north pole, the ice surface around Antarctica would be severely reduced, and the ocean would absorb more of the sun's heat.

Transitions from the liquid to the gaseous phase are critical in distillation, and it is through the process known as vaporization that those transitions are accomplished. Vaporization can occur by boiling and by evaporation, and the difference between the two is critical. Evaporation involves the phase transition of a compound that occurs below its boiling point at a specific pressure. It happens relatively slowly, and almost entirely at the surface of the liquid phase. In contrast, boiling occurs below the surface of the liquid, and proceeds much faster (Figure 5.6).

Heating up the liquid phase of a compound like water will, in general, increase the vapor pressure of the liquid. When the vapor pressure of the liquid is greater than or equal to the outside pressure, the liquid will begin to boil. The bubbles you see in a boiling pan of water are not oxygen or hydrogen being released. Rather, they are there because increases in temperature of the solution are accompanied by increases in the number of molecules that are transitioning to water vapor. Gaseous collections of water form in solution, and these physically push to the surface and rise out of the boiling solution. There is no chemical reaction happening here, but merely a phase change.

We know the transition temperatures from liquid to gas of many compounds.

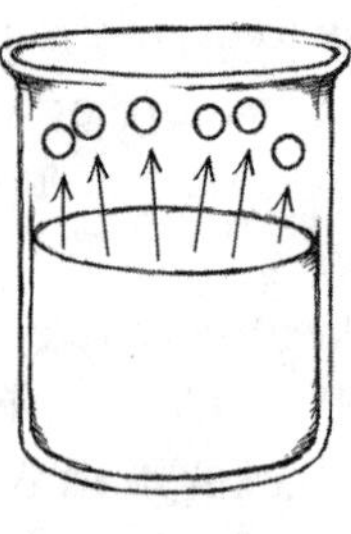

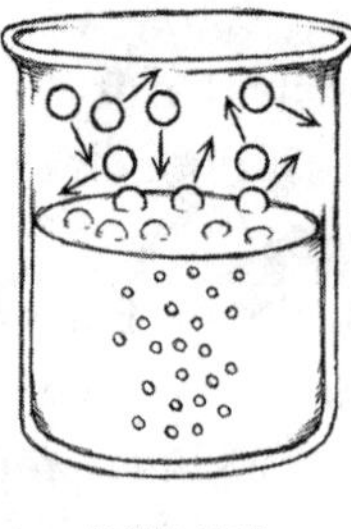

Figure 5.6. Evaporation versus boiling. Boiling is considered a "bulk phenomenon," and evaporation is called a "surface phenomenon."

Table 5.1. TRANSITION TEMPERATURES FOR SEVERAL COMPOUNDS, IN DEGREES FAHRENHEIT. MP STANDS FOR MELTING POINT; BP IS BOILING POINT.

COMPOUND	SOLID ⟷ MP	BP ⟷ GAS
Oxygen	−218	−182
Argon	−189	−186
Helium	−272	−269
Propane	−189	7
Methanol	−98	65
Ethanol	26	78
Water	0	100
Carbon	3550	3825

Table 5.1 shows several examples of the transition points from solid to liquid (the melting point) and liquid to gas (the boiling point). Because phase transitions can go two ways, the boiling point for water, for example, can be seen either as the temperature at which liquid water starts to become a gas, or as the temperature at which water vapor condenses into liquid water.

We have spent some time with water and its phase transitions not only because you can't understand distillation without knowing its characteristics, but also because this amazing liquid, so essential to life itself, is one of the most interesting compounds on our planet. It is one of the few common chemical compounds that can naturally be solid (ice), liquid (water), and gas (water vapor). Think about it. You can wake up on a late winter day with snow on the ground, and through the course of the day observe all three phases of water. The temperature rises well above freezing as the sun comes out and the solid snow melts to liquid water. As the sun heats the air up even more, the snow water vaporizes and rises as a gas. Curiously, fog and the steam from your shower are not considered gaseous. Think about that. Obviously, the shower water is not boiling, or it would scald you badly. As we said earlier, molecules are continually moving around and slamming into each other. When this happens to water at room temperature there is a transfer of kinetic energy, and some of the water molecules on the surface gain so much energy that they manage to escape the rest of their compadres and evaporate into the air. When heat is applied, the molecules start to move faster and faster and the evaporation of water molecules gets more intense. We know what happens when the water comes to a boil; and in that case, true gaseous water is emitted. But even when you are taking a steamy shower, what is emitted is liquid water (it has to be, because it hasn't reached the boiling temperature, which is where the phase transition occurs). Shower "steam," then, is just a bunch of liquid water molecules floating around in the bathroom, tiny droplets that are spread out so thinly as to appear steamy.

For distilling, it is critical to know not only the boiling point of water (100 degrees C), but also that of some other chemicals present in the liquid one wants to distill. If the object of distillation is to obtain ethanol by getting rid of water and any other impurities that accompany it in the mash, then the boiling points of ethanol and water are the need-to-know quantities. It just so happens that water boils at a higher temperature than ethanol (78.4 degrees C), and distillation proceeds by heating to below the boiling point of water, collecting the distillate, and discarding the stuff that doesn't vaporize. But the mash is a very complex stew. Besides containing things other than ethanol that the distiller wants to preserve, it will certainly contain things that he or she doesn't want. Some nasty extras in a whisky mash, or in a rough wine that is being distilled, likely include acetic acid, butanol, and propanol. These have boiling points at 118, 117, and 97 degrees C, respectively, all higher than that of ethanol. But some objectionable compounds, most notably the methanol that is terribly toxic to human nervous systems, actually have significantly lower boiling points (64.7 degrees C). Methanol is extremely hard on the human body, with effects that range from vomiting to unconsciousness to death. Ingesting this tiniest of alcohol molecules can also lead to blindness, because it is particularly toxic to the optic nerve. Ten milliliters of pure methanol are enough to blind someone, and a full shot glass will kill even the hardiest of spirit drinkers. Hence it is critical to distill away any of the methanol that will almost certainly be present in the mash. Fortunately, distillers can take advantage of that thirteen-degree C gap between the boiling points of methanol and alcohol.

As the mash heats up, but before distillers begin to collect the ethanol vapors coming off the mash, they will collect and discard the "heads"—any distillate formed at up to about 77–78 degrees C, or just before the ethanol starts to boil off. As a rule of thumb, for each ten-gallon batch of mash a distiller will pour off the first three-quarters of a pint, which will be mostly methanol and will smell like rubbing alcohol. Similarly, distillers also discard the "tails" that begin to be produced once the mash has heated significantly above ethanol's boiling point (and certainly by 95 degrees C). This part of the distillate tends to be rich in the fusel oils that give out an unwanted "wet dog" aroma. (Occasionally a distiller of Scotch whisky will want to leave in a few of those molecules, but never more than a trace.) In all cases, the water component of the mash will begin to boil while the ethanol is still vaporizing, so there will inevitably be some water in the "hearts" phase of the distillate that the distiller conserves. As a result, if you are using a traditional still you will have to make multiple passes before the alcohol concentration reaches the desired level. The distillate will also contain other compounds with boiling points close to that of ethanol, and where a highly pure spirit is required, distillers have had to come up with some ingenious techniques to remove those, too.

Many advances in the design of distillation equipment have been made in the almost two centuries since the invention of the column still. Out of necessity, industrial chemistry has produced some amazing variations on the column still, and spirits distillers have learned some tricks too. One innovation that has been important to top-of-the-line distillation is the reflux still.

A basic reflux apparatus is quite simple. As vapor is produced in the boiling pot and rises in the column, some of it condenses and returns to the boiler, while the rest is distilled off into the final product. This creates a cycle in which the repeated return of the condensed vapor in a reflux column both increases the alcohol content of the distillate, and allows an increasing fraction of water and other impurities to be removed. The key to this process is the presence of a fractionating column inside the reflux still that helps condense the rising vapor. Reflux stills tend to be very tall, because the taller the column, the better the reflux. Efficiency is also often improved by the use of a "boiling ball," a copper sphere that is placed where the distilled vapor passes in the column, and that increases the surface area on which condensation can take place. Figure 5.7 shows the major kinds of reflux still.

As the figure shows, there are three major kinds of reflux still: forced, external valved, and internal valved. Forced reflux columns, the most popular and best-known of the column stills, are renowned for their effectiveness, and use cooling

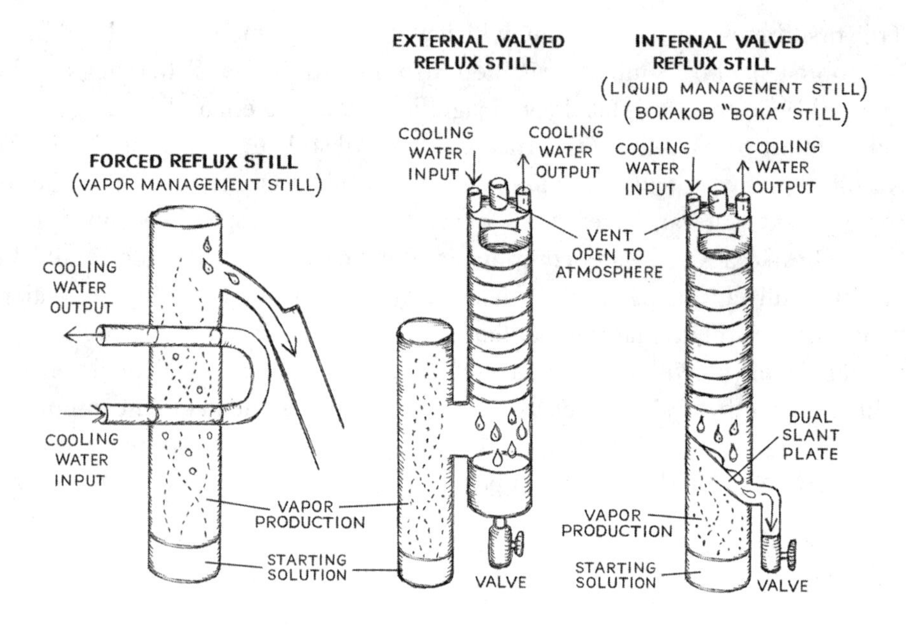

Figure 5.7. Types of reflux stills: forced, external valved, and internal valved.

tubes for the condensing. These cooling tubes (filled usually with water) run through the column in direct contact with the vapor, and the process is pretty straightforward: the vapor from the starting solution rises through the column and then flows into the cooling coil, where it condenses. Any vapor that does not come in contact with the coil makes its way through to the condensing chamber and is eventually collected. If the vapor hits a coil, it condenses immediately and falls back into the heated liquid before attempting to make it again through the cooling coil apparatus. This might happen several times to some of the solution in each distilling run, as the vapor constantly tries to overcome the obstacles that the cooling coils throw at it. The rate and efficiency of this process are, of course, controlled by the length of the cooling coil, and by the temperature of the water running through it.

The external valved reflux still performs the forced reflux step outside the heating column. As the figure shows, the reflux column is connected via a T-tube that extends from the main body of the vapor column. The reflux column usually consists of a cooling coil made of copper, a collection vat at the bottom of the column, and a vent that is open to the atmosphere. The valve allows the distiller to control what part and how much of the condensed vapor is collected. Whatever is collected is run repeatedly through the process, resulting in purer and more concentrated ethanol each time through.

The architecture of an internal valved reflux still is also shown in the figure. In this design, the forcing of the reflux happens inside the main column, and the collection reservoir is placed within the column. This reservoir is usually formed by placing a dual-slant plate somewhere in the column below the reflux apparatus. Both the internal and external reflux stills are popular with home distillers and are alternatively known as *liquid management stills.*

Stills are beautiful machines, and it is remarkable to contemplate how they have changed over time.

Now that we have some chemistry under our belts, let's follow a carbon atom through a modern distilling process. Our atom originated billions of years ago. In fact, most of the carbon atoms in your body are that old, with only around one in a trillion being younger, or new. Our token carbon atom, like most, was formed during that ancient time by induced fusion, after a nova had occurred and created the gas cloud that eventually coalesced to create our solar system. This rush of carbon was pretty much all that was needed to give our solar system enough carbon to end up the way it is. Our carbon atom likely floated around space until our planet Earth formed, four and half billion years ago, and then settled here.

Since then, this single atom has probably been incorporated into a lot of different living things. Perhaps they were plants, but more than likely they were bacteria, and it was passed back and forth via their metabolic systems as they lived and died. Eventually, our carbon atom formed a carbon dioxide molecule that floated around in the atmosphere until, probably a year or two ago, it found its way to a plant called *Vitis vinifera,* a.k.a. the grapevine. More precisely, it made its way into the pulp of a growing grape, to be converted to sugar by the photosynthesis process that was taking place in the chloroplast of a cell inside a grape produced by the grapevine.

The sugar molecule into which our carbon atom settled itself in the grape pulp had five carbons, arranged in a ring in a specific kind of sugar called glucose. There, it was one of the carbons in the sugar that got converted into ethanol. Remember that it might also have been in a position on the glucose that was converted into carbon dioxide, and simply have recycled back into the atmosphere. Once comfortably ensconced in the glucose in the grape's pulp, our atom was abruptly taken by a vineyard worker in Italy's Piedmont region to a winery. There the grape containing it was crushed, releasing the sugar molecule in which it was ensconced into the *must* (the freshly crushed grapes, also known as *pomace*). Once in the must, the sugar molecule was taken up by a yeast cell. There it became food for the tiny single-celled eukaryote, which used fermentation to produce energy from the sugar ring it resided in, and in the process created ethanol. Again, our carbon atom needed to be in the right position on the sugar ring in order for it to make its way to ethanol. Had it been in the wrong position, it would simply have been released as carbon dioxide, and recycled back into the atmosphere. We will resist the urge to overdo the chemistry, just noting that our carbon atom was lucky enough to make it into an ethanol molecule.

Once incorporated in that ethanol molecule, our carbon atom had to wait for all of the other sugar molecules in the solution to convert to ethanol. As it sat waiting, it might have noticed that smaller molecules like acetic acid and methanol were also swirling around it, as well as some relatively gigantic protein molecules that had made their way into the initial must. Some of those were large pigment molecules (the grape harboring our carbon atom was purple), and others were enzymes the grape had used to make energy and sugar. Fortunately for our carbon atom, the must was not used for fertilizer, or fed to animals after the crushing, but rather was sent to a distillery and used to produce grappa.

The juice from the crushing was drawn off to make wine, while the ethanol molecule our carbon was sitting in went to the distillery with the rest of the lightly crushed pomace. This stew of seeds, skins, stems, and any adhering pulp contained trillions of other carbons that had been incorporated into small molecules like acetic acid and methanol, along with a hoard of ethanol molecules that amounted to about 4 percent of the total volume. Unusually, grappa pomace has by law to be distilled without added water, which means that it needs to be moist and the initial distillation has to be gentle, over steam. The resulting *flemma* containing our carbon molecule contained about 15 percent ethanol when it was loaded into a column still. As the temperature around it rose, our carbon atom witnessed methanol molecules and other volatile materials rushing to the surface of the solution, then popping out and floating upward. These were the "heads," and they were destined to be discarded. At about 78 degrees C, the ethanol carrying our carbon atom began to join the rush toward the surface of the solution, saying goodbye to the water molecules, pigment molecules, and other larger enzymes and proteins it was leaving behind. From the surface it traveled, as ethanol vapor, to the top of the column; but because it was in a reflux still, it condensed and fell back into the solution. It then made a couple of other false starts up the reflux column before eventually making its way to the condensing column and being collected into a large steel vat. At that point, it saw only water and other ethanol molecules around it, because everything else had either been distilled off earlier, or left behind in the boiler (and by everything, we mean most of the water, those pesky pigment molecules, and all the other enzymes from the grape and the yeast).

At the end of the distillation part of its trip, our carbon atom was content to rest in its position as one of the ethanol molecule's two carbons. The local ethanol concentration was by then about 80 percent, so it had lots of company. Together with its companions, it stayed in its steel tank for at least six months, and likely more, before it was diluted (flooded with water to an ethanol concentration of about 40 percent) and bottled—with an intermediate stop in an oak barrel if the grappa batch was destined to be *invecchiata* (aged). Once the bottle was purchased and drunk, our atom was digested along with the rest of the grappa, and its host molecule was dismantled—leaving our carbon atom to float around once more in the environment, awaiting the next stage of its unending journey.

6

To Age or Not to Age?

We knew that maturing spirits passively in wood barrels significantly changes them. But what if a bourbon were to experience the mellowing effects of wood aging as well as the temperature changes and sometimes violent agitation involved in a sea voyage? This kind of extreme treatment had, after all, made the fortified wines of Madeira famous in the sixteenth century, and England's India pale ales equally renowned in the eighteenth. What would a similar treatment do to an American whiskey? The makers of this bourbon had tested that out, and had liked the result enough that our sample was from their Voyage 17, which had turned out to be pretty smooth sailing. This uneventful history may have accounted for the relative unassertiveness of the whiskey itself: medium amber in color, and with wet tree bark on the nose, it was smooth and unctuous with notes of orange peel and vanilla overlying its caramel base. It lingered on the lips and at the back of the palate, and we thought, maybe imagined, that we could detect the briny effects of the sea air. We had no way of knowing what it had been like to start with, but it had completed its long ocean voyage damn well.

Anyone who has been in this world for a while knows that advancing age has its drawbacks. But whereas for humans this unpalatable truth is both undeniable and inescapable, in a world where a bottle of fifty-two-year-old Macallan can go for fifty thousand dollars, it's pretty clear that the passing of time is a boon to spirits. Or at least to some spirits. To those for whom the qualities of the base ingredient (the agave in a white tequila, for instance, or the fruit in an eau-de-vie) are paramount, aging by itself adds nothing to the essence; and indeed, resting one of those lively spirits for a long time in an oak barrel might well be anathema to a purist. It is also generally agreed that neutral spirits, whether straight from the still or rectified with botanicals, do not as a rule benefit from aging—though that hasn't stopped some modern producers from producing "barrel-aged" or "barrel-rested" gins (actually, a historical holdover from the days when barrels were pretty much the only bulk storage available).

If you are a grappa aficionado, however, barrel aging is an option you'll definitely want to consider. For although unaged grappas can be wonderfully delicate, the most prized bottlings have usually spent some time in wood. And if your preferences run toward corn or rye whiskey, you will almost certainly insist on some barrel age. Not without reason is unaged whiskey usually derided as moonshine, though even this typically—though not necessarily—rough beverage has its own enthusiasts (see Chapter 20). Continuing along the spectrum, if you are a Cognac producer, your product is obliged by law to spend at least two years in French oak, just as any Scotch whisky, to be considered worthy of the name, has to pass at least three years (and a day) in barrel before being sold. Both of these fabled spirits often go far beyond these minimums, of course; and most of the more expensive spirits you will see in your local bar will likely have spent some time in wood. (See Chapter 12 for a whisky-maker's perspective on the subject.)

So when it comes to spirits, age can be very important. But don't put too much stock in it. At the end of 2018, BBC Scotland reported that precise measurements of age, made by using a highly sensitive variant of the radiocarbon dating technique, had shown that twenty-one of fifty-five bottles of old and "rare" Scotch whisky were either outright fakes, or not from the year declared. Among them, ten purportedly nineteenth-century single malts proved fraudulent. Given that the dated bottles had been gathered as a random sample from auctions, retailers, and private cellars across the United Kingdom, collectors of old and expensive whiskies were rattled to the core, though those familiar with the shenanigans of the collectible-wine market may have not been too surprised. The organization that commissioned the study concluded that some $55 million worth of fake rare Scotches was in circulation at the time on the secondary market, dwarfing the estimated $48 million value of all the whisky consigned to auction across the United Kingdom in 2018. The organization's founder declared that, while most vendors were not knowingly selling fake Scotch whiskies, every allegedly rare bottle, especially if a single malt, "should be assumed to be fake until proven genuine." Caveat emptor.

If, despite the hazards, you prefer your spirits aged, the key item of interest to you will not be the simple passage of time, but the nature and origin of the wood in which the aging was done. When it emerges from the still, any new spirit will be pretty stable, and if it is kept in a full, unreactive, and impermeable container it will remain the way it is more or less indefinitely. But if it is stored instead in a wooden barrel, several things will happen. First, the spirit will take up flavors from the barrel in the form of molecules that were present in the wood to begin with, molecules derived from charring or toasting (see below), and, if the container was previously used for wine or another spirit, residues left by the barrel's previous occupant. At the same time, the slightly permeable wood will permit both a tiny amount of oxygen to diffuse into the barrel from the outside air, and a corresponding proportion of its contents to evaporate. As all this gradually proceeds, the additional molecules from the wood will enhance the chemical complexity of the liquid and thus its potential range of flavors, while the slight oxidation resulting from diffusion will often increase the intensity of those flavors and produce a variety of new compounds. With continuing evaporation, the level of liquid in the barrel will go down. And while a winemaker will almost always keep his or her barrels topped off to reduce the space for oxygen inside them, a spirits-maker usually will not, instead simply watching the volume of liquid reduce as the "angels' share" of the spirit disappears into the atmosphere and the oxidation process accelerates. Together, these various processes sustain chemical reactions that over time will fundamentally alter the flavor, aroma, and color characteristics of the liquid inside the barrel. By controlling for barrel type, size, age, and length and conditions of maturation, the cellarmaster can thus aim for a particular gustatory result—though, since the aging of spirits is as much an art as it is a science, an element of unpredictability remains.

There are many types of wood. And several tree species (maple, chestnut, mulberry, and hickory among them) are occasionally used in barrel-making. But in the European tradition from which today's major spirits categories sprang, the wood of the oak tree has long been preferred by barrel-makers, mostly because of its unusually wide "radial rays," the woody bands that give individual staves the strength necessary to maintain the barrel's integrity. There are many species of oak, but only a few are optimal for cooperage, and nearly all maturation of spirits is done in barrels made from a small handful of oak species, most commonly the European pedunculate oak, *Quercus robur,* and the American white oak, *Quercus alba.* The seasoned barrel wood derived from both species "breathes" via tiny channels through which nutrients had moved when the tree was alive, and liquids inside the barrel are prevented from seeping out through the empty channels by the presence of tiny knots known as "tyloses" that largely block them (Figure 6.1). Pedunculate oak from Europe has fewer tyloses than white oak, and what this means for the cooper (as opposed to the distiller) is that pedunculate barrel staves have to be split along the grain of the wood, rather than more conventionally sawn, making European barrels more time-consuming and thus more expensive to manufacture.

From the distiller's point of view, though, oak from both sides of the Atlantic has advantages that go far beyond the physical characteristics that make it the barrel-maker's friend. Every kind of wood is chemically complex, but oak contains a particularly desirable combination of chemicals while lacking normally unwanted compounds such as the resins that are present in pine wood. Molecules in oak that winemakers and distillers can take advantage of to add complexity and color to their wines and spirits include the lignins that impart vanilla flavors and spicy undertones, the tannins that contribute astringency and texture, the alde-

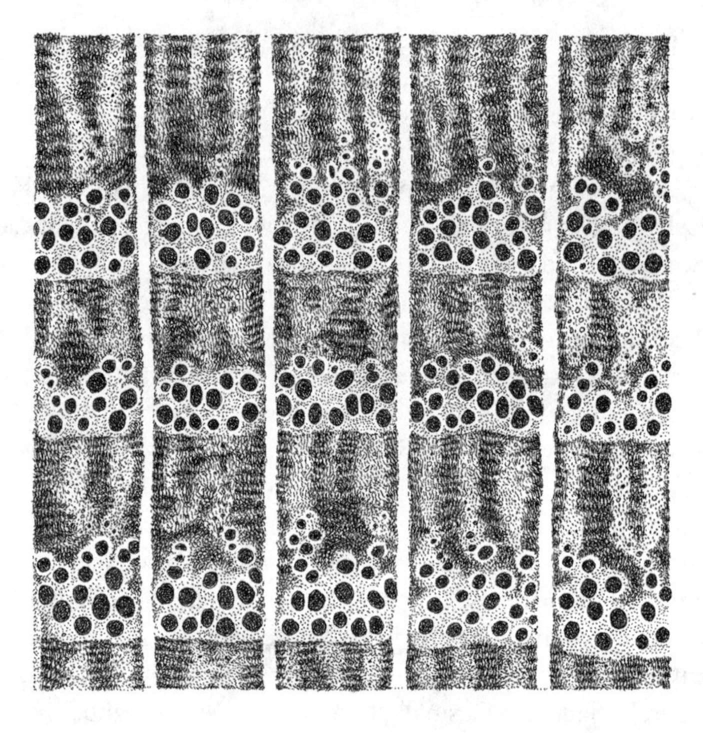

Figure 6.1. Magnified cross-section of a plank of American white oak showing the tyloses, which are the dark dots inside the lighter-colored xylem channels.

hydes such as furfural that give bready or caramelly notes, and the lactones (derived from lipids that are particularly abundant in American oak) that offer woody aromas and sometimes also a whiff of coconut or clove. Traditionally, coopers making bourbon barrels bent their staves over fire, igniting and charring the barrel interiors in the process. Such charring (which is a little different from the "toasting" of wine barrels that involves longer periods at lower temperatures) turned out to invigorate the lignins, enhancing their vanilla flavors, and to caramelize some of the natural hemicellulose sugars in the wood, producing toffee and nutty flavors that vary in intensity in approximate proportion to the amount of char applied. The burned interior barrel surfaces also act somewhat like charcoal filters, trapping unwanted compounds such as the sulfides that can impart an unpleasant metallic taste.

With their fewer tyloses, French oak barrels (obligatory for aging Cognac) are more permeable than their American counterparts (obligatory for bourbon), and the wood from which they are made is softer and more finely grained. Together, these physical characteristics give the European barrels faster rates of micro-oxygenation than their American equivalents, generally resulting in shorter aging times in this respect. In terms of chemical influence, European oak barrels are prized for the high polyphenol levels that yield spicy notes after charring. Those polyphenols also include the tannins that contribute astringency and structure. American oak, in contrast, is lower in phenols but higher in the aromatics, such as vanillin, that contribute sweetish and toasty notes during aging, and it is particularly high in those woody, coconutty lactones. As for specific aromas, you'll often hear the words coconut, smokiness, coffee, and cocoa mentioned in connection with American barrels, while for European wood the most common descriptors include honey, vanilla, dried fruits, nuttiness, and spices of various kinds.

At this point we should emphasize that *Quercus robur* and *Quercus alba* are not quite the only oaks in the business. Particularly with the recent blossoming of the Japanese whisky industry, we are hearing more about Mizunara oak (*Quercus mongolica*), a Japanese native that was initially pressed into service as a cask material during shortages that followed World War II, but has since achieved a cult following. It is prized for its very high vanillin levels, as well as for imparting sandalwood-spicy flavors and a whiff of incense to the whiskies that are aged in it. And then there is the sessile oak, *Quercus petraea* (a.k.a. *Q. sessiflora*), a European and Near Eastern native that is used more widely for timber than for barrels, but is commonly used in aging Cognacs because it is the main oak growing in the famous Tronçais and Limousin forests of central France—the only two legal sources of wood for aging Cognacs. The sessile oak is said generally to contribute less structure but somewhat more aroma to wines and spirits aged in it than the pedunculate oak does; and it has its own regional peculiarities, varying from place to place in its tannin levels and its sweet and spicy components. This species has recently attracted particular attention in Hungary, where a cooperage industry based on it has lately expanded significantly.

But precisely because every species is variable in this way, just knowing the oak species at issue tells only a small part of the story when it comes to aging spirits. European barrel oak is identified by geographical origin, with each major forest (there are six of them in France) having its own traditional consumers, and its own local peculiarities that include the mix of species. Across the Atlantic, some makers of brown spirits prefer white oak from the Ozarks, while others promote the virtues of Appalachian white oak, or wood from the cool upper Midwest where slow growth results in a tighter grain. The differences between regional varieties occur largely because the qualities of the wood are very responsive to local growing conditions that include not only climate, but also topography, altitude, and exposure. Within a particular species, slower growth—a general characteristic of European *Quercus robur*—will tend to promote higher levels of many flavor components. And how the wood is treated after harvesting is also important: the properties of the resulting barrels are affected, for example, by how and for how long the wood is dried.

Because there are so many variables, barrel choice in the aging of spirits is a crucial yet bewilderingly complicated business. Few authorities have had the temerity to generalize much in this area—and indeed, much barrel knowledge is proprietary, and closely held. But a couple of decades ago two French experts, Pascal Chatonnet and Denis Dubourdieu, baldly declared that "European sessile oak and American white oak are perfectly suitable for aging fine wine [whereas] pedunculate oak, with its low aromatic potential and high ellagitannin content, is best suited to aging spirits." Whether they had a good technical or scientific point can be argued forever, but it is clear that, probably because of the heavy weight of tradition, nobody on either side of the Atlantic was listening.

Whatever the source of the wood, spending time in any oak barrel, even briefly, irreversibly changes a spirit. Whether that is the same as improving it is entirely a subjective judgment, although there will certainly be more flavor components in the barrel-aged spirit than in its unaged counterpart. Technically, the barrel-aged product should also be darker than the unaged one, although any clear spirit can also be darkened by using additives, usually a caramel colorant. Conversely, some barrel-aged light rums and other spirits are filtered to remove the color that was gained from the oak. Adding to the complexity is the obvious fact that not all barrels are equal. Oaks from different sources offer different combinations of compounds, while different degrees of char also produce different flavor profiles. And new barrels naturally have higher concentrations of compounds than older ones do, because some of the barrel's finite store of molecules is given over with every use. How many times a barrel is reused is a very subjective decision, unless it is specifically governed by law. At least one winemaker in Burgundy uses his brand-new barrels for maturing his cheaper second wine, because he finds their first-time effects overpowering. Only in later years does he use those barrels for his subtler and more expensive single-vineyard wines. In contrast, bourbon-makers are required by law to use only brand-new charred white oak barrels, and as soon as the whiskey it contained is bottled, a barrel is sold. Often the buyer is a maker of Scotch, whose own product will inevitably be influenced by whatever had been in barrel before. This fact offers Scotch-makers significant new flavor possibilities, which is why some Scotch whiskies may reside in multiple successive barrels as they age, for example starting out in bourbon barrels before finishing their maturation in a sherry cask. Not using new barrels suits Scotch-makers well, not only because old barrels are less expensive than new ones, but also because the flavors in a whisky made from barley are more delicate than those in a whisky made from corn, and are thus more likely to be overwhelmed than complemented by the influence of new wood. But while used bourbon barrels traditionally went to the makers of other spirits, used barrels of everything from whisky to Madeira to rum have become appealing to brewers looking for new ways in which to flavor their beers. After about three successive uses, the average wine or spirits barrel will effectively have yielded all of its original compounds. But it will still retain its ability to breathe, and unless it leaks or is otherwise unusable, a used barrel is always valuable.

The size of the barrel in which a spirit matures is another significant variable in the aging process. A larger barrel will offer a smaller surface area-to-volume ratio and will thus yield fewer wood compounds and less oxygen diffusion per unit volume than a smaller one. While wine barrels vary enormously in size (with wine barrels used in warmer climates tending to be bigger and thus more thermally stable when full), spirits-makers are less concerned about the environment of maturation and generally prefer their barrels on the smaller side. The fifty-three-gallon (two-hundred-liter) bourbon barrel is a typical example, and one Hudson Valley whiskey producer has reportedly used barrels as small as ten liters.

In principle, spirits should age faster both in newer barrels and in smaller ones. This is, indeed, the main argument for using small barrels, although the result doesn't exactly mimic the results of longer aging in larger ones. Importantly, though, it turns out that how fast a spirit matures is also heavily dependent on local climate. Specifically, maturation rate is very responsive to both humidity and temperature. A barrelful of spirits sitting in a warehouse in Jamaica will mature much faster than one in cooler Scotland, while one stored in Kentucky will fall somewhere in between. This is partly because the seasonal drop in temperature in the north will slow the maturation process almost to a stop, whereas in the more equable Caribbean it will happily burble along year-round. Maturation rate is also partly a function of the average ambient warmth: many bourbon aging warehouses ("rickhouses") are sited and constructed to maximize the amount of heat they absorb during the summer. In this, bourbon-makers differ from their Scottish counterparts, who prefer cooler cellars. It is also relevant that greater swings in temperature will affect the expansion and contraction of the spirits in barrel, forcing the liquid into and out of the wood and thereby increasing extraction from it. Local humidity is a factor, too, affecting the rate of diffusion of fluid through the barrel walls and hence the oxidation process. For these reasons and more, spirits produced in northern climes typically have longer maturation times than those made in warmer places. Until recently, it was not at all unusual to encounter a Scotch that had been aged for twelve or eighteen years, whereas bourbons in that age category were traditionally much harder to find. Indeed, the first twenty-year-old bourbon was marketed only in 1994, although as the price of a twenty-three-year-old bottle of Pappy Van Winkle tops five thousand dollars on the secondary market, we are already seeing an increase in the number of older American whiskeys. Farther south, under Mexican rules a tequila is considered "reposado" after a mere two months to a year in cask, and "extra añejo" after only three years. Still, in this case it seems that agave-based spirits inherently age faster than those based on cereals: tequilas tend to lose their sweetness after about four or five years in barrel, ultimately becoming (perhaps excessively) dominated by the wood flavors derived from the cask.

Distillers also have a large bag of tricks for artificially speeding up aging, or for mimicking the effects of barrel age. Thus, in addition to the caramel that is often used to darken younger spirits, it is legal for those who produce Cognac, the world's most esteemed brandy, to add tiny quantities of an astringent oak extract known as *boisé* to their product during aging. Appropriately enough, the boisé itself is often aged. A newer technique is to imprint the interior barrel surfaces with a honeycomb pattern that increases the effective surface area of the wood that is exposed to the liquid. Wood barrel inserts have also been used to speed aging, with one trendsetter producing results almost overnight by switching from positive to negative pressure and back to force spirits into and out of fresh wood. Alternatively, at least one American company is currently offering a proprietary micro-oxygenating technology to help speed up the aging process in spirits that include whiskies and, remarkably enough, moonshines. And don't forget Endless West, the maker of "molecular spirits" made "overnight" to rival the "finest aged whiskeys" (see Chapter 23). Still, imitation is the sincerest form of flattery; and such subterfuges only emphasize that, where the depth and complexity of a fully mature spirit is desired, traditional barrel aging is the gold standard.

There is basically no legal requirement, anywhere, for the makers of a spirit to print the age of their product on the bottle label (see below for an exception). In western France's Charente region, where they make Cognac, the law in fact expressly forbids distillers to do so. Wisely enough, since most are blends from different years, each Cognac is instead awarded a quality grade. Significantly, though, those grades are related to maturation time: the youngest component of a VS Cognac has to have at least two years in barrel; that of a VSOP at least four years, that of a Napoleon at least six, and that of an XO at least ten. The only category that depends strictly on a quality judgment is the "Hors d'Age" rating that producers exceptionally apply to their finest Cognacs—although of course a quality judgment is also at least somewhat implicit in the geographic denomination (Grande Champagne, Borderies, Bois à Terroirs, and so forth) that every Cognac bottle aspiring to top quality must bear. The rules in the Charente, however, contrast significantly with those in Gascony, the region to its south. There, the distillers of Armagnac often proudly post vintages when single-year products are involved, although the VS (youngest component two years in wood), VSOP (four years), and XO (ten years) classifications are also used for the blends that account for most Armagnac production.

In the United States, the recent explosive growth of whisky sales has led to a substantial depletion of older stocks, so much so that despite the recent popularity of single-cask bottlings, which are by definition from a single year, it is now rare to see the age of a spirit featured on the bottle. Most U.S. spirits on the market, too, are still blends, many made on an enormous scale with the contents of many individual barrels being mixed in a large vat in pursuit of the blender's particular vision, or of an established house style. Whether the aim is to attain an imaginative new result, or a predictable product that will meet the expectations of long-term customers, such blending is a high art that often requires balancing different qualities of the spirit that emerge during successive stages of the aging process. Hence the predominance of multiple-year blending. As in France, any age statement on a bottle of American spirits blended in this way must reflect the age of the youngest element in the blend. That is, according to the federal regulations, "age may be understated but shall not be overstated," so that a bourbon labeled as eight years old may contain some twelve-year-old spirit, but not vice versa.

It actually is a bit more complicated than this. A spirit labeled *straight whiskey* must have spent two years in new oak. If it spent between two and four years in that new oak, then that fact has to be spelled out. Anything labeled *bottled in bond* will be a blend of different spirits with at least four years in cask, and all from the same year. And *single cask* means just what it says. Otherwise, though, all bets are off. So be careful when scanning the label of any spirit, and above all don't be fooled into thinking that any prominent number it features necessarily refers to the age of what is in the bottle. It may simply be an arbitrary number used to designate the particular bottling and have nothing whatever to do with the spirit's age.

So, aged or unaged? Well, it depends principally on the spirit, and on the consumer's palate. Pretty much everyone would agree that grape brandies and whiskies are the spirits that benefit most from barrel aging, while the notion that rums and some tequilas also respond well to a bit of time in cask is hardly controversial. Most of us would equally agree that such beverages as eaux-de-vie, in which the critically important thing is to capture the essence of the base ingredient, are often best bottled straight from the still. That leaves us, of course, with an immense middle ground in which a spirit will undeniably change in wood, but not necessarily improve. Here we are purely in the realm of individual taste. Our own bottom line is that we would hate to live in a world in which, say, all grappas or tequilas were aged, or none were. One of the most delightful aspects of spirits is their sheer variety, a variety that is wonderfully amplified by the magic of oak.

7

Spirits Trees

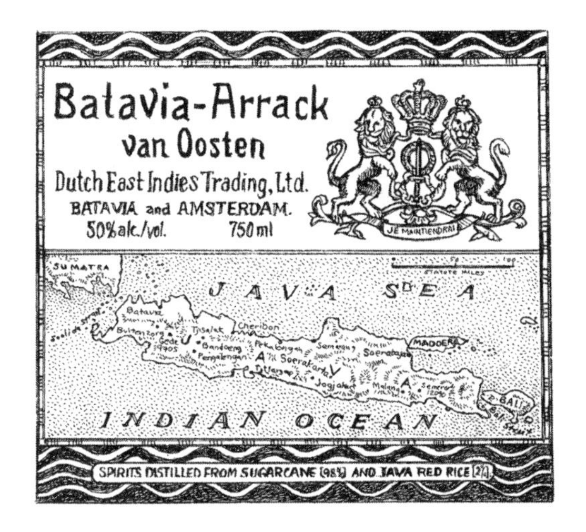

For this chapter we just had to try the Grand Daddy of all the New World spirits: arrack originated on the early Portuguese plantations in the Canary Islands, and so may predate cachaça or rum. Sadly, we could not find one made by distilling palm toddy, but the green bottle in front of us held a direct descendant of the Canaries spirit, pot-distilled in Java from sugarcane and with a touch of red rice in the mix. We didn't know how much of the funkiness typical of rice alcohols such as baijiu to expect; fortunately, after a lightly fragrant nose, it was the sweetness of the sugar that dominated. Despite its hefty 50 percent ABV, this pale, unaged spirit was remarkably unctuous and smooth on the palate, with an oddly fruity finish. It was a very decent sipper, even though its early precursors had been most famous as the basis of tropical punches, and its producer evidently regards it as a mixer, printing punch and cocktail recipes on the back label.

Readers of our book *A Natural History of Beer* will know that, as systematists both by profession and inclination, we like to try to understand the relationships among the things we write about. Here we identify how the various types of spirits relate to each other on the basis of their characteristics. This is not easy, because there are so many different characteristics to consider, and their relative importance and relationships are a matter of opinion. Systematists working on living animals don't, by the way, have this problem. This is because evolution in the biological sphere proceeds by splitting lineages, so it is pretty clear that what systematists need to know is how the organisms they study are related by descent. In contrast, what we try to do here, for better or worse, is apply to spirits the methods used by biologists to determine the relationships among living species, using as many agreed criteria as we can find. Whether this is a productive exercise is for the reader to judge, but we believe it will be at least thought-provoking.

Aristotle was probably the first person to explain in writing why the endeavor of classifying living things is important. He even introduced an idea for how to systematize life. He called it the *scala naturae,* the "ladder of life," and it was based on a scale of perfection. The more perfect you are, the higher you are up the ladder. This conveniently allowed Aristotle to place humans at the top of the scale, while other organisms appeared lower down according to how far from perfect they appeared to be. Thus birds were a bit below mammals, snakes were yet lower, and worms were near the bottom. Obviously, any criterion can be used to arrange things on a scale like this, and alcoholic spirits did not evolve in the same sense that life has. Nonetheless, inanimate things like spirits do have a history, and this history has usually resulted in their possession of certain character-

Figure 7.1. A *scala naturae* of spirits on a t-shirt. Imagine the following arrangement of bottles, *from left:* gin, rum, tequila, plum brandy, rye whisky, Scotch, baijiu, peach brandy, schnapps, and grappa.

istics. As a result, comparing those characteristics is a good way to establish the relationships among them. There are many ways to present those relationships, most of them visual. Intuitively, for example, general proximity in a diagram is a good way to display relationships, and Figure 7.1 shows one t-shirt designer's attempt to do this. It is essentially a *scala naturae* based on the typical bottle sizes and shapes of various alcohols.

Although this way of systematizing things is visually appealing, it doesn't help us much in thinking about how these alcohols relate to one another. For instance, on the t-shirt a plum brandy is the bottle fourth from the left, and a peach brandy appears far away, a full five bottles over. But any rational grouping of these spirits would show the two brandies' close relationship by placing them very near each other in the diagram. Given the degree of ambiguity that the *scala naturae* allowed, it may be surprising that it survived until Charles Darwin's time, morphing, thanks to the efforts of medieval scholars, into a religious interpretation (angels added above people, God above the angels). In addition, it nurtured one of the more misleading notions in natural history: special creation. Without an alternative way to think about how the natural world had diversified, early natural history had been stuck with the notion of deliberate creation, so that when evolution came on the scene, it was predictable that lingering notions of the ladder would give rise to the misconception that one contemporary form had morphed into another, as in the idea that chimps evolved into humans. We now know that chimps and humans had a common ancestor that lived about seven million years ago, and was neither chimp nor human. But in the absence of fossils of that ancestor, we can only estimate what it was like by comparing the features of humans and chimpanzees.

The story of how this ladder-like way of thinking was debunked and replaced is a long one that we don't need to go into here. Suffice it to say that in the mid-eighteenth century the Swedish savant Carolus Linnaeus invented the binominal (two-name) nomenclature that we use today for living forms (for example, we are *Homo sapiens,* belonging to the species *sapiens* of the genus *Homo,* which contains several other species, all now extinct). This opened the way to the idea that life is organized into an entire hierarchy of categories that contain smaller categories. A century later, Charles Darwin added the idea of evolution: "descent with modification."

One of the most interesting things that Linnaeus did was to establish a hierarchical "key" to the identification and classification of living things. This device was a forerunner of what is today known as a *decision tree,* and it involves reasoning very similar to that used in Figure 7.2, a diagram designed to help you decide which brand of whisky you would be most pleased with.

By using this diagram and answering a series of yes or no questions, it is easy to reach a conclusion as to which kind of whisky one might prefer. For example, if you answer yes to the following questions, then you will prefer the Bunnahabhain 18:

> Have you had whiskey before?
> Did you enjoy your taste of whisky?
> Was it a peated, smoky whisky?

By contrast, you will prefer the Yamazaki DR brand if you answer no to these questions:

> Have you had whisky before?
> Are you a vodka person?
> Are you a beer drinker?
> Do you have balls of steel?

Again, this is very much like what a taxonomist would call a taxonomic key (Figure 7.3).

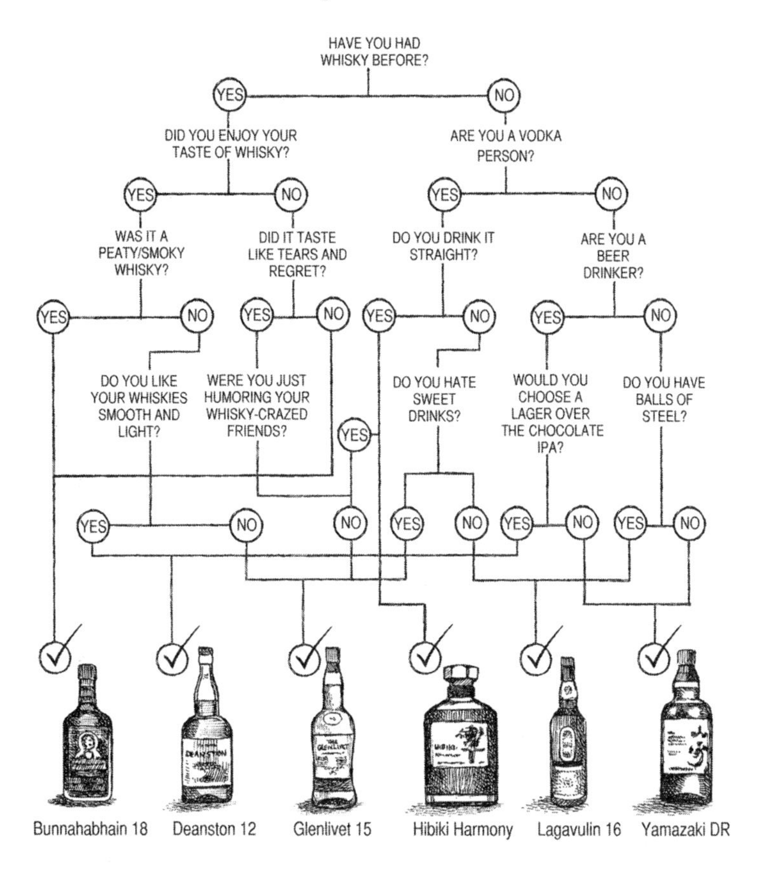

Figure 7.2. A decision tree for choosing a whisky, redrawn after *Bootleg Brew: The Bootleggers Guide to Choosing the Perfect Dram;* Instagram, Bootleg Brew, November 19, 2015, https://www.instagram.com/p/-QnVvdwk0F/?taken-by=bootlegbrew.

In this case, the organism is a member of the Angiospermophyta (flowering plants) if the answers to the following questions are all yes:

Is it a plant?
Does it have roots?
Does it have seeds?
Does it have flowers?

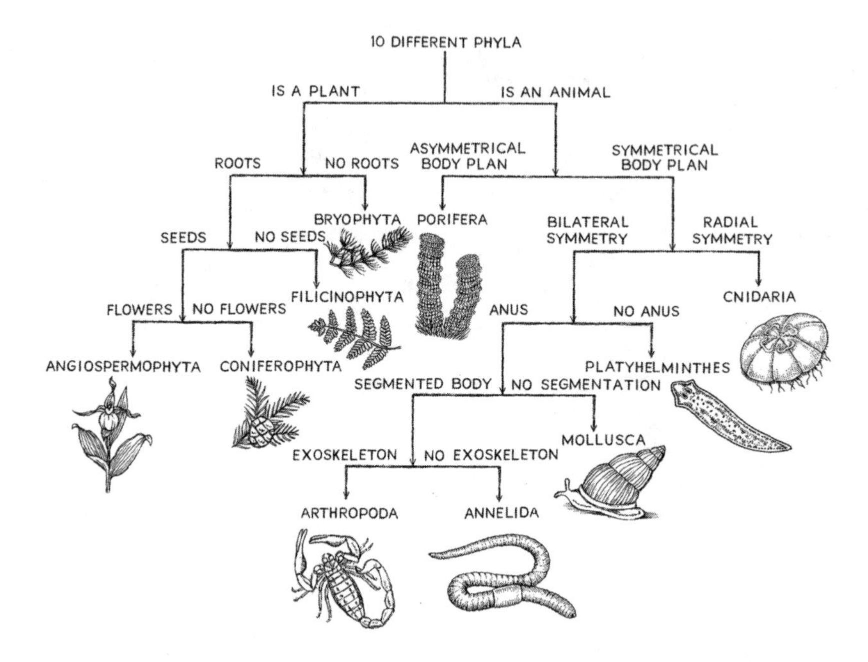

Figure 7.3. Taxonomic key for plants and animals.

For his part, Darwin coined the term *tree of life* to describe the diversity of organisms on Earth as a great branching tree, in which all life has a common ancestor at its root. This way of thinking came to dominate biology in the early twentieth century, but many of the trees drawn in this period were based purely on the authority and expertise of the scientist who drew them. It wasn't until the 1960s that sound scientific techniques were developed to reconstruct common ancestry, creating trees that represented scientific data and that could be converted to hierarchical classifications. These are the techniques we will draw on here.

There has actually been a lot of speculation about the relationships among beverages in general, and among spirits specifically, although the resulting hierarchies are based on authoritarian interpretations like those in early twentieth-century systematics. Many readers may have seen such diagrams without recognizing that they are networks of relationships of the kind also seen in evolutionary trees. Interestingly, the abundance of whisk(e)y trees out there makes it appear

Figure 7.4. A Venn diagram of whiskies.

that whisk(e)y distillers and drinkers are incredibly interested in the genealogy of their favorite beverage, while other spirits have elicited less genealogical interest.

Figure 7.4 shows the simplest whisk(e)y diagram we could find: a Venn diagram of the kind used to impart hierarchy to a set of data. The figure implies that all rye whiskies fall firmly into one group, all bourbons in a second group, and all Scotches in a third group. More importantly, those three kinds of spirits together form a single group (represented by the outer circle) called whiskey. A more complete treatment of all spirits using a diagram like this would include gins in a single circle, vodkas in another, and so on. Some trees of whiskies end up on t-shirts, like the famous one shown in Figure 7.5.

This network actually shows the wonderful hierarchy that has developed around whiskies. Note that the different steps in the hierarchical arrangement are represented by different geometric shapes. For instance, the major divisions of whiskies into American, Irish, Canadian, and Scotch varieties are depicted by rectangles. The next organizing factor is specific to each of these four groups: American whiskey, for instance, is organized around whether it is corn-based, wheat-based, bourbon, single malt, rye, or blended. Each of these subgroups is represented by a hexagon. Spirits such as American bourbon can be divided even further (by the dashed hexagons) into Tennessee and Kentucky bourbons. At the tips of the hierarchical network are the individual distilleries that fit into the groups to which they are connected by lines. For instance, Jack Daniels comes from the Tennessee Bourbon hexagon, which comes from the bourbon hexagon, which comes from the American whiskey rectangle, which is part of the general

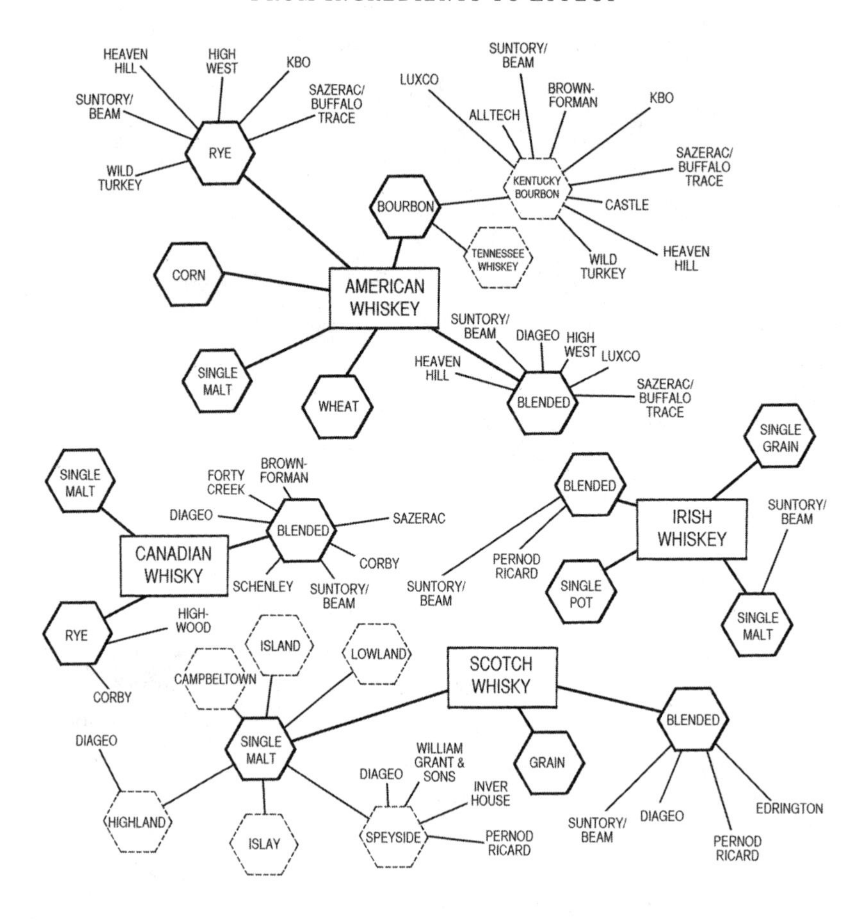

Figure 7.5. A network diagram of whiskies.

spirit called whisky. If you look really hard at the diagram you will see that there is no other hierarchy in it. In other words, Scotch isn't more closely related to any other whisky than it is to all the rest. Thus each of the four major whiskies—American, Irish, Scotch, and Canadian—has its own separate tree.

Figure 7.6 illustrates a similar diagram, this one by Jason Haynes. This "tree" takes the whiskey classification a bit further by adding Finnish, English, Japanese, German, African, Indian, Welsh, and Australian varieties. Note that, once again, the whisk(e)ys from different countries are not arranged hierarchically but rather form their own networks. In biological phylogenetics, this sort of arrangement is called a star phylogeny, and it would indicate that all of the major whisky

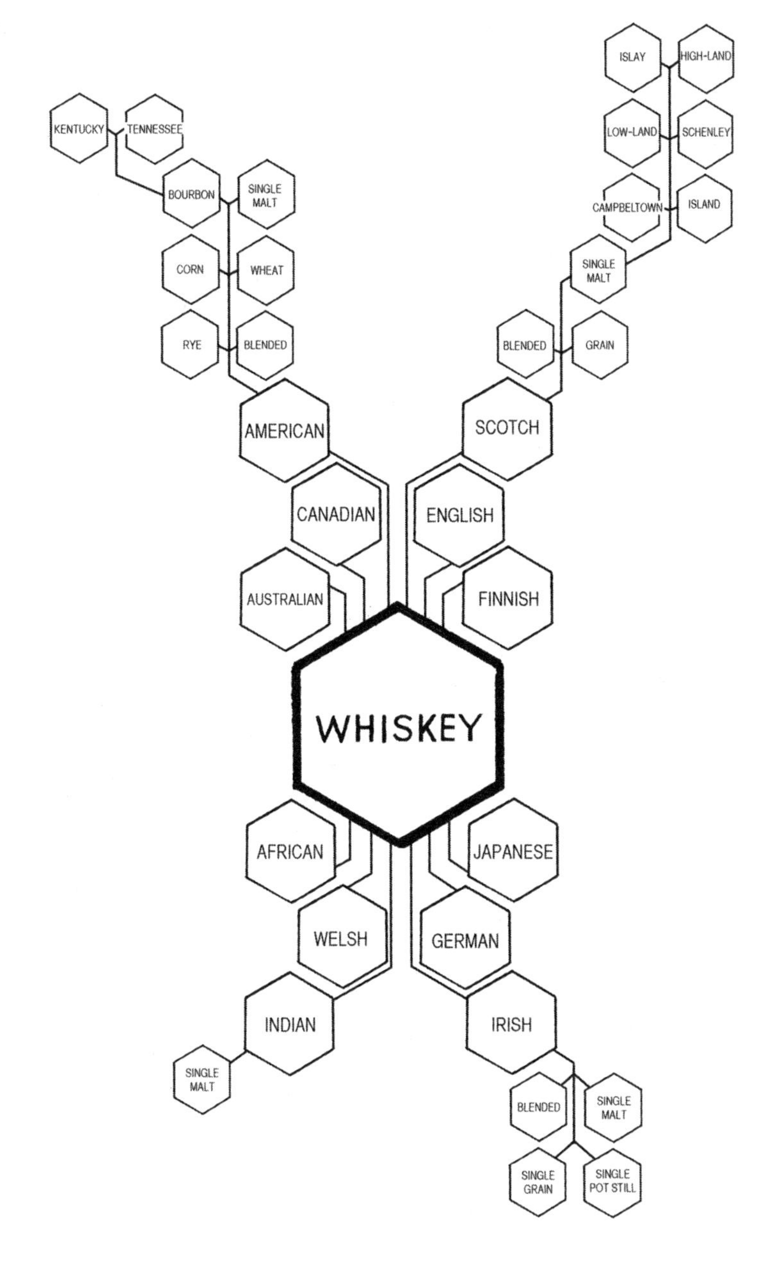

Figure 7.6. Whiskey diagram by Jason Haynes, https://www.pinterest.com/pin/522065781780667251.

branches emerged independently, or that they all diverged so quickly that it is difficult to tease apart the relationships of, say, Irish to Scotch to English whiskeys. By contrast, this diagram does show some branching *within* Scotch, American, Irish, and Indian whiskeys.

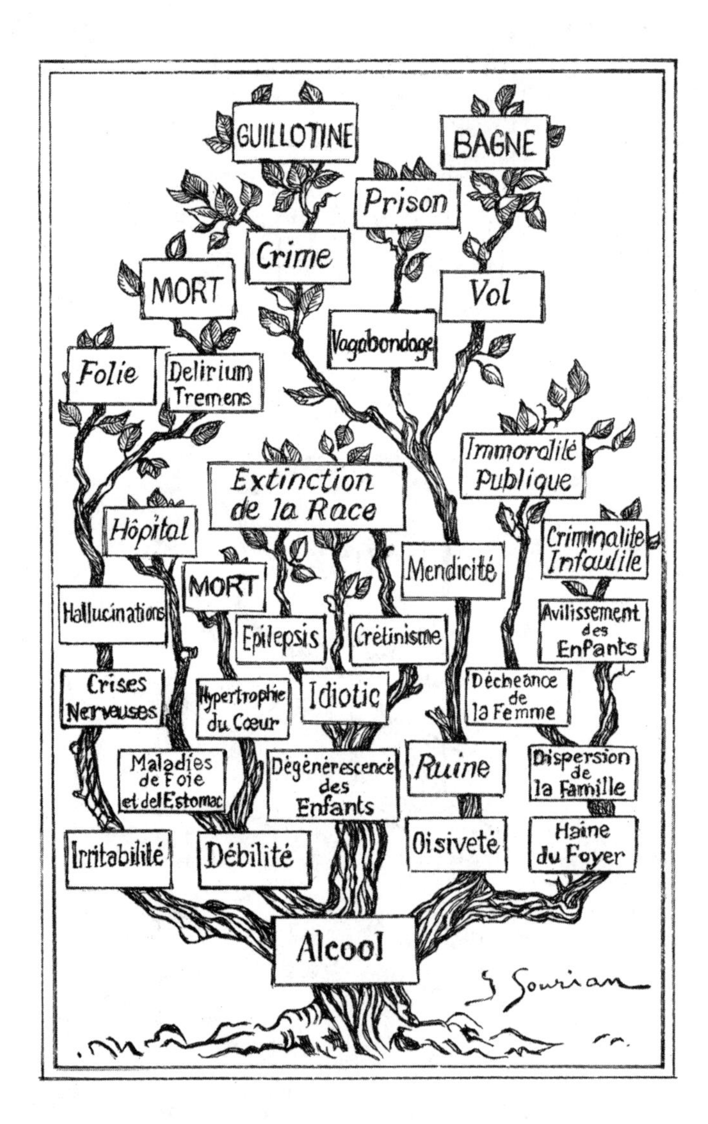

Figure 7.7. A tree depicting the woes of alcohol consumption. French circa 1900.

Some of the diagrams are not subtle with respect to their tree-like implications. A rather nice graphic by Chris Rassiccia shows a tree of whisk(e)ys that, like the t-shirt network, has four great branches: Irish, Canadian, Scotch, and American. Not subtle, but it does show how the various tips of the tree are derived by branching from common ancestors. One amusing variant of the classic biological tree is the temperance tree in Figure 7.7. This French postcard from circa 1900 shows the dangers of imbibing alcohol, specifically absinthe, and implies that drinking alcohol will inevitably lead to disease or the guillotine, or even to the extinction of humanity.

After that rather depressing reflection, let's return to trees that represent spirits. Figure 7.8 shows another whiskey family tree with the same four major kinds of whiskey (the big limbs coming from the trunk of the tree—Irish, Scotch, bourbon, and rye). What makes this tree different, though, is that it has data associated with it. By formal definition, Irish whiskey is "Aged in wood casks at least three years. Made in Ireland." Scotch whisky "Contains malted barley. Aged in oak casks at least three years. Made in Scotland." Canadian whisky "Contains some rye. Aged at least three years." And so on. These bits of information are what a systematist calls "characters" or, in this case (forgive us), "synapomorphies," suggesting that, for example, containing malted barley, aging in oak barrels at least three years, and manufacture in Scotland are characteristics that all Scotch whiskeys have, and that no other kind of whiskey has. This makes them analogous to the answers to the questions in the taxonomic keys we showed in Figures 7.2 and 7.3.

If we step back a bit, we can demonstrate how such characters can be used to build a tree. Say we have four items we want to analyze: wine, brandy, beer, and whiskey. There are only two obvious characters we need in order to construct the tree: whether the beverage is made from grain, and whether the beverage is distilled or not. We will also add flavored soda to our analysis, not because it has either of these characters, but because we need this fifth item to "root" our tree, that is, give it a direction. Systematists would call this fifth item an "outgroup" (the four other items are called "ingroups"), and they would refer to the five beverages as "taxa." See our simple matrix of the data which lists the items in a column, and the characteristics in a row:

	GRAIN	DISTILLED
Soda	No	No
Wine	No	No
Brandy	No	Yes
Beer	Yes	No
Whisk(e)y	Yes	Yes

Figure 7.8. Bearings Guide whiskey tree. Redrawn from www.seekpng.com.

We can then transcribe this into a presence-absence matrix, in which zero means "absent" and one means "present." So we have:

	GRAIN	DISTILLED
Soda	0	0
Wine	0	0
Brandy	0	1
Beer	1	0
Whisk(e)y	1	1

The trick here is to find the tree that best fits the character states. There are fifteen different trees in which wine, brandy, beer, and whiskey (with soda as an outgroup) can be arranged. Since the 1960s, various researchers have developed ways to approach arranging them. The simplest of these use "Occam's razor," or the principle of parsimony, proposed by the English fourteenth-century scholar William of Ockham, who declared that we should prefer the simplest explanation for anything. The parsimony approach accordingly judges as best the tree with the fewest number of changes needed to accommodate the data. Let's look at one of the fifteen trees to see how this works.

The tree that wins the parsimony test is the one that puts beer with brandy as each other's closest relative, while wine and whiskey are grouped together. Figure 7.9 shows how our two characters might have evolved if the above groupings prove true (a process called character mapping). For the characters to fit into the evolutionary scenario shown in the figure, we need to identify two events on branches involving grain, and two on branches that involve distillation, for a total of four steps. When we look at another of the fifteen trees, the one that puts beer with whiskey, and brandy with wine, we obtain the evolutionary scenario in Figure 7.10.

Because in this tree topology one of the characters (grain) can be explained by an event in the common ancestor of beer and whiskey, while the other character, distillation, has two events, the total here is three events. This tree is clearly more parsimonious than the first one, with four. If none of the others required only three changes, then the tree in Figure 7.10 would be accepted as best. If only things were so simple! There turns out to be one other tree of the fifteen that is also three steps long. It is shown in Figure 7.11.

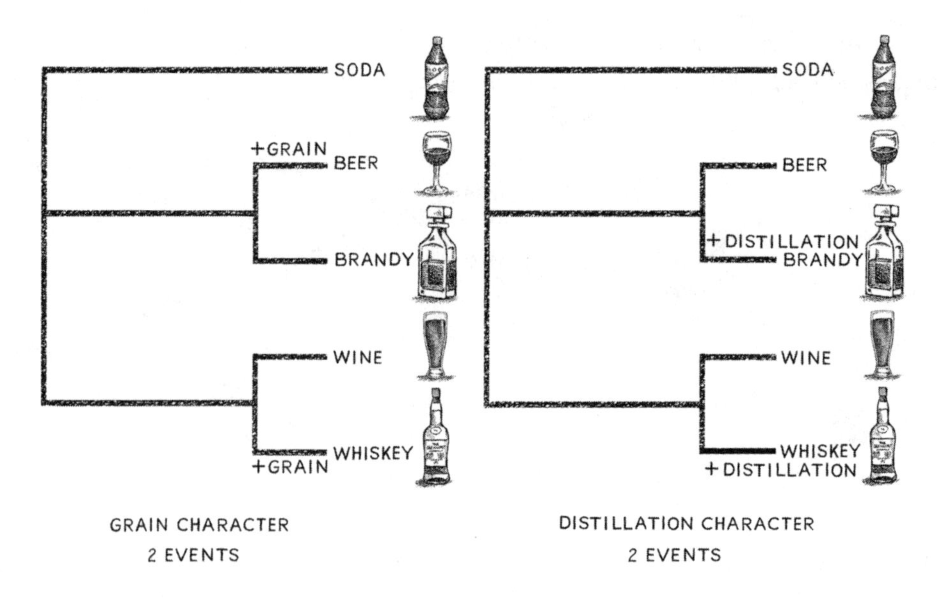

Figure 7.9. Mapping beverage characters on the wine-whiskey tree.

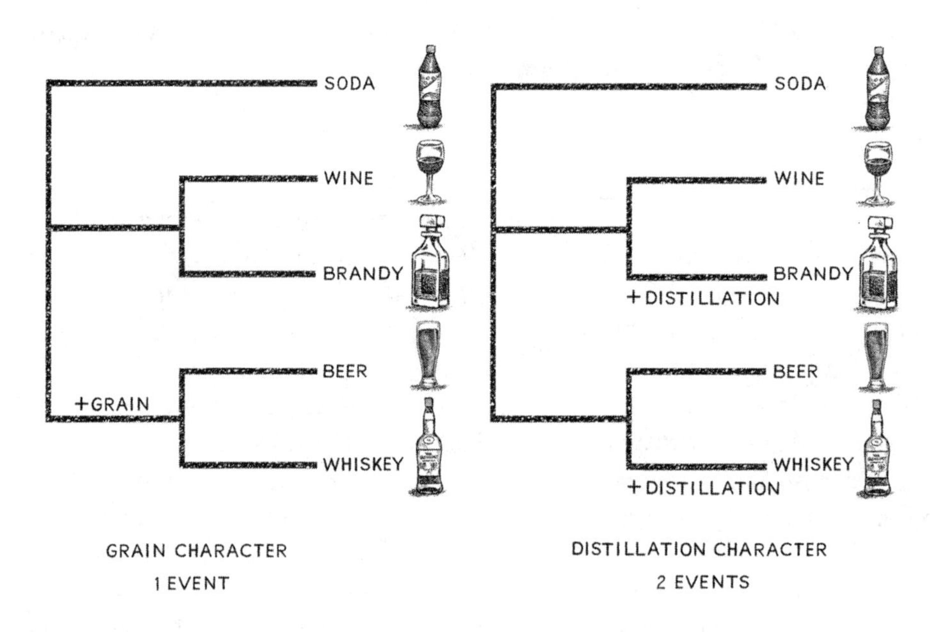

Figure 7.10. Mapping beverage characters on the beer-whiskey tree.

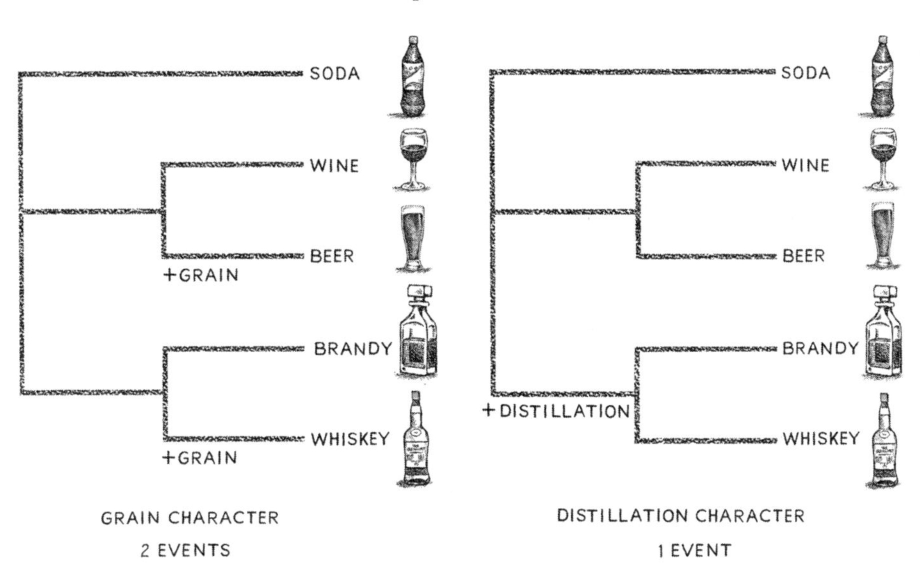

Figure 7.11. Mapping beverage characters alternately on the brandy-whiskey tree.

The tree topology in Figure 7.11 is thus just as parsimonious as the topology in Figure 7.10. So, are we stuck? Fortunately, no. If we worked only in the realm of parsimony, we would need to add more features of all the beverages to help us make a decision. But there are other ways to proceed on the basis of the data we have, because many systematists have argued that there is no reason to think that parsimony is the way that evolution always works. Instead, they create "models" of the evolutionary process, and use them to estimate probabilities of change. In our example, this means that we need to ask what else we know about these beverages that suggests that one tree topology has a higher probability than another.

Our model might, for example, put a higher probability on the choice of plants as raw material than on the distillation conditions. If it does, we can use an approach called "maximum likelihood" that involves estimating probabilities for all of the possible fifteen trees, then choosing the one that meets this criterion. The result of likelihood analysis using our model is a single topology that is the same as in Figure 7.10 and has a likelihood of −11.19034. The equally parsimonious tree in Figure 7.11 has a lower likelihood: −11.86640. All of the other thirteen possible trees have likelihoods of between −11.86644 and −12.54254.

For those of you who are tree-building aficionados, the way we proceeded was to use a DNA-based model that incorporated a generalized-time-reversal (GTR)

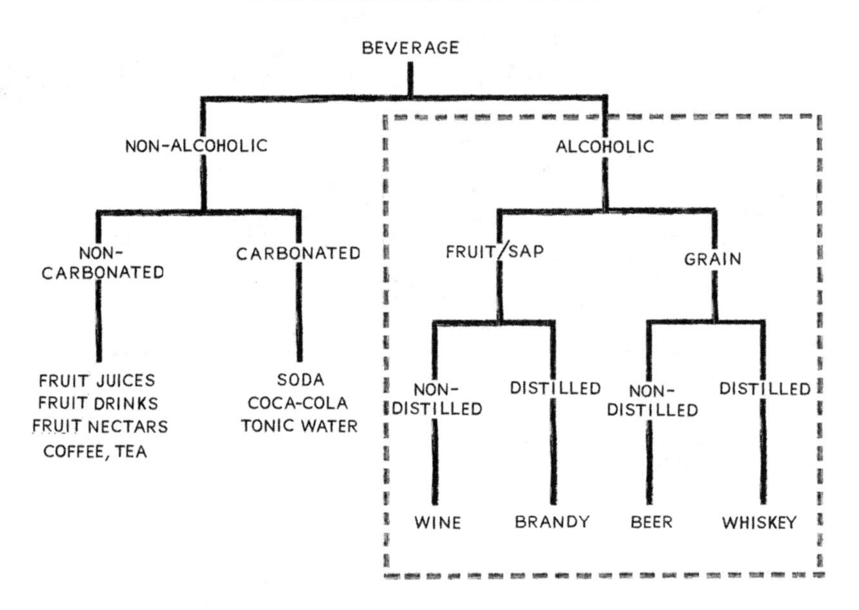

Figure 7.12. Tree of beverages in general. The tree is used to explain how characters are collected and analyzed in a phylogenetic context.

model with a gamma distribution and the shape parameter estimated from the data. The GTR model simply allows any of the four bases in the DNA sequence, namely guanine, adenine, thymine, and cytosine (G, A, T, and C), to change into any of the four possibilities, and the gamma distribution is just a parameter to estimate the rate of change of the sequences. To get this to work we needed to change the zeros and ones in our data set into As and Ts, respectively, for the grain character, and into Cs and Gs, respectively, for the distillation character. We reasoned on the basis of probability that an appropriate model would be to prefer the distillation state over the plant used. We then used a program called PAUP written by our colleague David Swofford to do the likelihood analysis. While the analysis we have done here might seem frivolous, it suggests what might be done if better data were available.

Peter Fellows presents the diagram in Figure 7.12 for beverages in general. It is like a decision tree, but we can use it for comparison with our results. Fellows included fruit juices and soda in his analysis, but we can disregard those "taxa" to the left and just focus on the right-hand side of his tree. The area in the dashed box is identical to the most parsimonious topology in Figure 7.10. But remember that the topology in Figure 7.11 was just as parsimonious as the one in Figure 7.10, and if you look at Fellows's tree closely, it is clear that he obtained it only because he used the "grain versus fruit" criterion as his first decision fork, and distillation as his second. If he had reversed the forks on the decision tree, he would have obtained the same tree as in Figure 7.11 (the tree that had lower likelihood than Figure 7.10). So essentially, what we did in our analysis was to prefer a model that placed the greatest likelihood on Fellows's first fork—the choice of plant material.

A single step better in the parsimony analysis, and a probability difference of 0.67610 in the likelihood one, are hardly large margins; but in modern phylogenetic analysis, huge numbers of characters are typically used, not just two. And trees constructed from data that are relevant to distilling do actually exist. This is because most of the yeast strains that are used for wine making and beer brewing, or for fermenting mashes for distillation, have had their genomes sequenced. These genetic blueprints can be used to provide characters for phylogenetic analysis, and researchers have used them to generate the family tree of the yeasts used for fermentation.

An interesting example of tree use in distilling comes from Zheng Li and Kenneth Suslick at the University of Illinois. These researchers invented an optoelectronic "nose" that can characterize spirits. They analyzed sixteen different spirits: four with high proof (including a 116 proof Willett Kentucky Single Barrel), six with medium proof (including an 86 proof Evan Williams Black Label), and four with relatively low proof (including an 80 proof Glenfiddich). Li and Suslick then took the data and made a branching diagram (a dendrogram using yet another method of diagramming, but we'll spare you the details). For an outgroup they used several grain ethanols diluted to about 50 percent, with their results shown in the tree in Figure 7.13. The isoelectric nose was able to group all of the high-alcohol beverages together, all of the medium proof beverages together, and all of the lower proof spirits together. This was a nice demonstration of the utility of an objective tool in identifying alcohol levels. But the nose sorted only by alcohol strength—and while the researchers used only 0.1 milliliter of each spirit in the analysis, they did not say what they did with the leftovers.

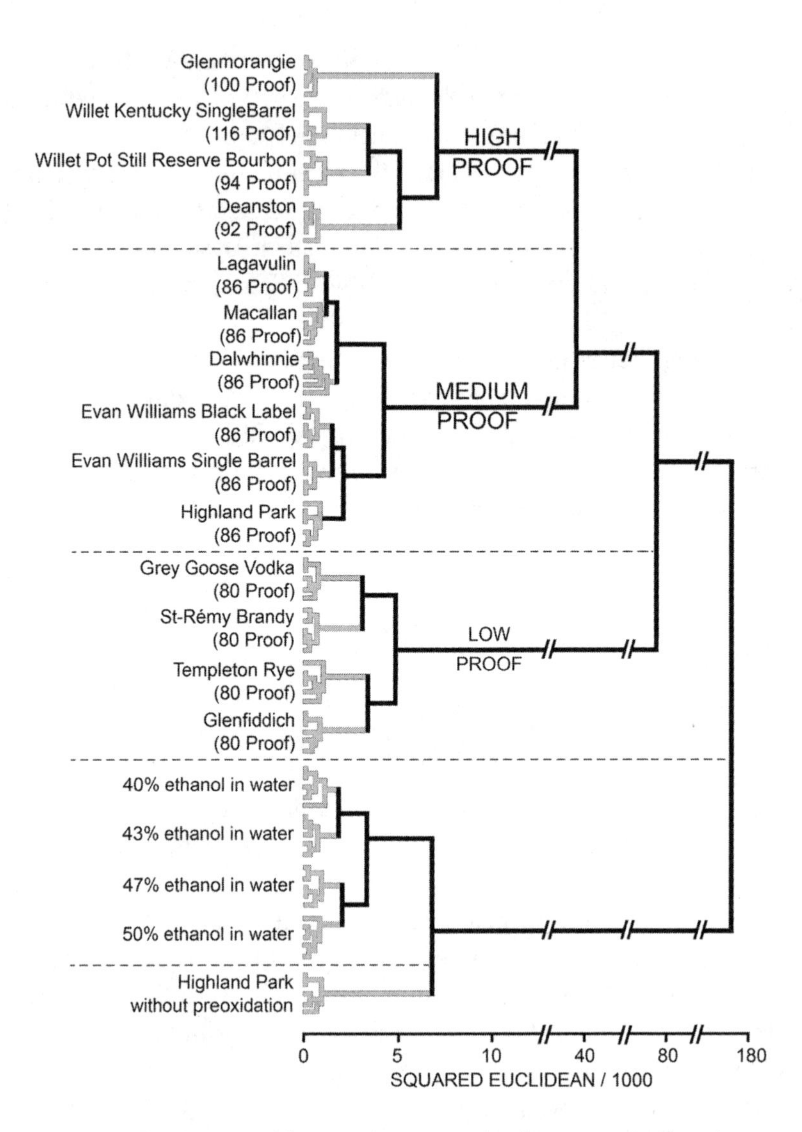

Figure 7.13. Dendrogram of fourteen liquor samples, four aqueous ethanol controls, and one spirit control treated differently from the low-, medium-, and high-proof beverages.

Our literature search failed to turn up a single tree or genealogy that included all major spirits categories, so we thought we should present our own tree of booze here, in part to show more clearly how such trees are made. The first step in any phylogenetic study is to collect your organisms, or taxa. In this case, the collection was made from our collective liquor cabinets and a little browsing in our neighborhoods. We then compiled a list of forty spirits that represent all of the "big six" (gin, vodka, rum, whisky/whiskey, brandy, and tequila), as well as one eau-de-vie. Once we had all of our taxa, we proceeded to research each kind of spirit for characteristics that might be useful in a phylogenetic analysis. We came up with almost thirty that could be used to generate a tree.

The first few characteristics simply define each spirit by one or two well-established traits. These include (1) uses juniper (yes, no); (2) uses potatoes (yes, no); (3) uses agave (yes, no); (4) uses grapes (yes, no); and so on. Other characters involved the common proof of the spirit, its country of origin, whether the spirit was infused or sweetened, whether it is usually aged, and if so, in what kind of barrel. Next we placed the data into a matrix much like the two-character ones we presented earlier. We then fed the data into two programs that analyze data by parsimony—Tree analysis using New Technology (TNT), by Pablo Goloboff and colleagues; and Phylogenetic Analysis Using Parsimony (PAUP), by David Swofford. We attempted to include an outgroup like beer or wine, but that destabilized the tree, probably because these drinks are so fundamentally different from spirits. Without those outliers, the two programs gave the same answer, which we diagram in Figure 7.14. As systematists we are impressed by this tree, because it gives us inferences that make firm sense. Believe us, some data sets do not behave nearly as nicely as the one we have generated here.

Note that this tree closely clusters the six major kinds of spirits into *monophyletic groups*. This clumsy term is a very important organizing principle in systematics, and it means that their members all share synapomorphies, that is, characteristics that indicate descent from a single common ancestor. The dots on the network in Figure 7.14 indicate where in the network the synapomorphies lie that define the "big six." For instance, dot 1 refers to the common ancestor of all brandies, defined by coming from distilled wine; dot 2 refers to the common ancestor of rum, defined by cane sugar alcohol as the source of distillate; and so on.

Some surprising results include the clustering of whisky and one example of eau-de-vie, and the connection between gin and rum. Some of these connections, however, rely on the location of the root of the network. For instance, if the tree is rooted on the eaux-de-vie lineage (which is not such a bad idea, although

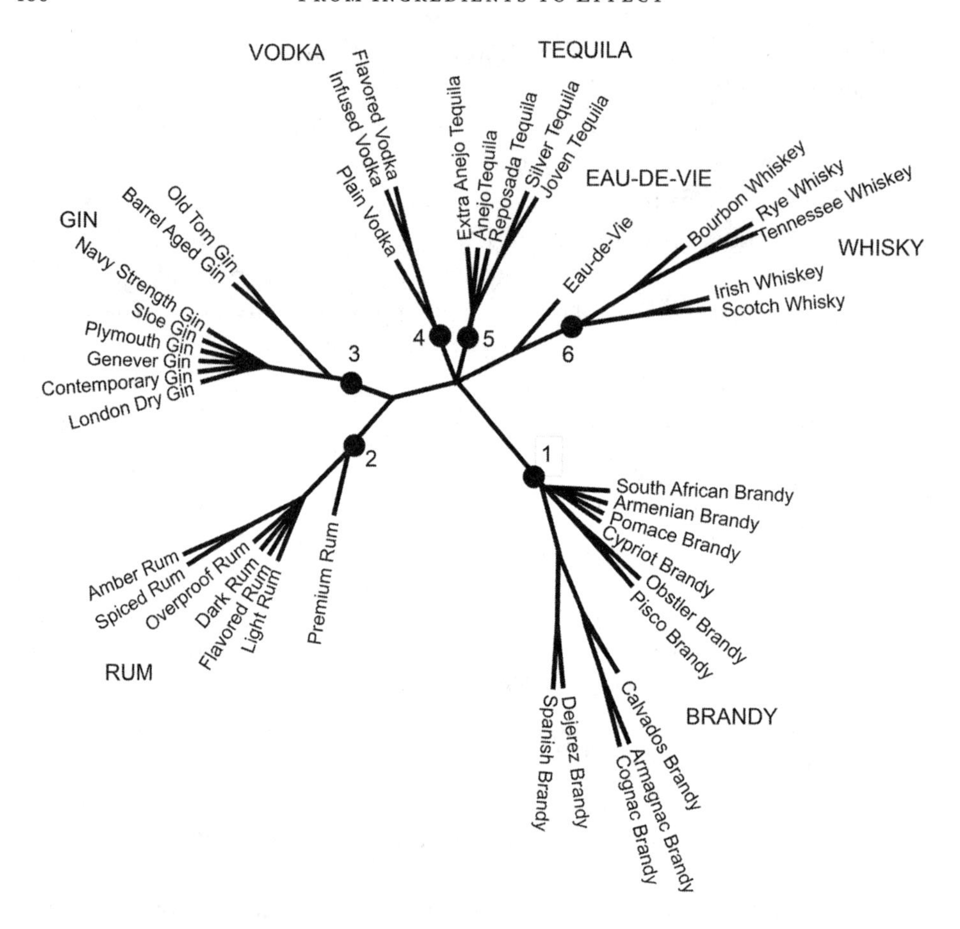

Figure 7.14. Phylogenetic unrooted network of the major kinds of spirits. The tree was generated with parsimony, using twenty-eight characters coded from common descriptions of the spirits. The characters included country of origin, plant material used, aging, and infusion.

we wish the sample size were bigger), then the inference that it clusters with whisky disappears. The relationships within the major monophyletic spirits are also interesting. For instance, within the brandies the French brandies cluster together, as do the Spanish brandies, and these two groups are themselves joined, indicating the close relationship between them. Pisco and Obstlers oddly cluster together too, while the rest of the fruit brandies are unresolved within the group. By looking at which taxa branch first within the monophyletic groups of spirits, we can try to make inferences about which spirit in each group is the most primitive (closest to the ancestral form). It appears that premium rum is the first rum to branch off in that group, plain vodka is the first vodka to branch off, while the Old Tom and barrel-aged gins pair together and branch off first. Obviously, it is debatable whether those particular within-group inferences mean much in historical, as opposed to technical, terms.

Given our very small and informal sample, there is not a lot more we can write at this point, except to remark how neatly the major groups of spirits fell out in our analysis: they sorted much more cleanly, for example, than beers did in an earlier assay. This may partly reflect the nature of distillation, in which unwanted compounds are removed wholesale from the spirits late in the process. Brewers, in contrast, achieve their aim by being selective with yeasts and other ingredients at the very beginning. It additionally seems possible that, from the very outset, distillers have typically shared a specific ideal for the type of spirit they produce, encouraging ideas of purity and tradition. This also contrasts with brewing, which has a much longer history, and whose practitioners, especially during the recent craft beer revolution, have been both more experimental and more protective of their proprietary yeast strains. But perhaps most of all, the star-like form of the diagram may reflect the speed with which distillation spread in Europe during the late thirteenth and fourteenth centuries. Local distillation traditions were established early on, rapidly became culturally ingrained, and retained their distinctiveness thereafter. Among the "big six" the only true newcomers are rum and tequila, a clear reflection of the unique histories of these distinctive spirits and the exotic sugar sources from which they are derived.

A caution: our database is not only limited in breadth, but is also composed of representative examples; and, as anyone who has tasted a lot of spirits knows, it can be difficult to discriminate even among major categories. For example, on occasion even an experienced taster may find it dismayingly easy to confuse a cereal-based spirit with a cane-based one, especially if significant barrel time was involved along the way. Discriminating among different wines is actually a lot easier, since the best of them will tend to express both their varietal character and their terroir. But while spirits may also vary significantly in the overall quality and provenance of their ingredients, distillers seldom if ever specify the precise varieties of the base constituents of their products, or exactly where they were grown (a minor exception: various high-end fruit-based products such as some marcs, grappas, and eaux-de-vie). In the future this may change, at least at the top end of the market; some craft distillers are already beginning to pay more attention to the exact sources of their ingredients, with interesting implications. Meanwhile, although it is maybe a bit unsettling for systematists, many spirits drinkers will continue to enjoy the element of mystery that almost every bottle presents.

8

Spirits and Your Senses

The liquor glowed promisingly green in the bottle, and although it was made in New York City's historic Harlem neighborhood, it stood squarely in the tradition of the absinthes that had accompanied the downfalls of such nineteenth-century European luminaries as Vincent van Gogh and Paul Verlaine. Not quite sure what to expect, we buried our noses in our glasses and immediately felt as if transported into a fresh spice garden. The aromas of anise, fennel, and the obligatory wormwood exploded into the headspace, along with a surprisingly strong suggestion of cucumber. All of these components continued seamlessly and deliciously along the palate, and on to the finish. Traditionally, absinthes are drunk with a hefty addition of water, sugar, and ritual; and at a lofty 66 percent ABV, it seemed like a good idea to follow custom in this case. This time, however, those extras simply dulled the liveliness of the spirit—which didn't turn cloudy, even at high dilutions. Still, the experience made us feel that Verlaine might have been on to something.

From the sound of clinking shot glasses of bourbon to the first smell of a peaty Scotch, from the glinting of a barrel-aged rum to the texture and taste of an infused vodka, spirits were made for our senses. And what is it that interprets all of the sensations that we appreciate in spirits? In the end, it is our brains, which can also be influenced in other ways such as inebriation and, unfortunately, addiction.

Our brains are complex, to say the least. And they are not finely tuned machines perfected by millions of years of evolution: instead they are jury-rigged contraptions, with amazing parts and properties accumulated over a vast period of time. So why would millions of years of evolution make our brains so untidy? Well, evolution does not take a straight path to perfection, as many of us were taught to think. Many of the brain's characteristics are neither perfect, nor inevitable, but rather came about as immediate solutions to problems posed by environmental challenges. Since evolution doesn't start over every time one lineage diverges from another, organisms are loaded with traits that they have simply because their ancestors had them. In other words, once a lineage attains a particular solution to an evolutionary problem, it can't simply discard it, go back to the start, and try again. Nor can it conjure up a new solution out of thin air. Change has to proceed on the basis of variations that already exist. Random factors interject themselves as well, through a phenomenon called drift that is very prominent in small populations (like those typical of our human ancestors). No wonder our colleague Gary Marcus calls our brains *kluges* (clumsy, inelegant solutions that work nonetheless). But the good news is that our brains happen to be pretty good at processing and synthesizing the amazing array of sensations we get when we drink spirits.

If you sit down with Laphroaig Lore, a very peaty, smoky Scotch, your eyes will tell you that the bottle is green-tinted, and that the label is a plain drab green. The green of the bottle obscures the color of the Scotch inside, but we will get to that in a moment. Your eyes detect that the bottle is cylindrical, and on a table about two feet from you. At this point, you can't smell, hear, taste, or touch any part of the spirit itself (you can touch the bottle, but there is nothing special about that). And while detecting the color doesn't sound like using lot of brain power, this initial sighting of the bottle actually uses both a very complex collection apparatus (the eyes), and a sophisticated processing system (the brain).

That bottle of Laphroaig is a solid object, and it is being bombarded with light. The actual nature of light is somewhat contentious. We will not venture a guess as to what light is; but what we do know is that it acts like both a particle and a wave. As the light passes through the bottle and the spirit within, some of it is absorbed and some is reflected, and the many wavelengths of light reflected from the bottle are what our eyes "see." These wavelengths are from across the spectrum of electromagnetic radiation, from tiny gamma rays (smaller than the diameter of a cell's nucleus) to gigantic radio waves (equal to the distance from New York City to Detroit). Most of the sighted organisms on this planet have keyed in on what is called "visible light," a small sliver of wavelengths that are about the length of a groove on a compact disc (400 to 700 nm), although some organisms have evolved ways of seeing in the near infrared (800 to 2,500 nm) and ultraviolet (100 to 400 nm) regions. Photons of light in its particle state can collide with things, hence the reflection and absorbance of light as it hits the bottle.

Our color detection system starts up when light reflected (or emanating) from what we are looking at enters the eyes. The human eye has several layers, but the most important for our Scotch drinker is the structure at the back called the retina. The retina is made up of millions of specialized cells known as rods and cones. These are arranged like a field of randomly growing corn stalks, although the rods are found mostly on the periphery of the retina and the cone cells are scattered throughout the rest, with a high concentration in a central region called the fovea. Both kinds of cells have what are called pigment proteins embedded in their membranes, and it is with these that light from the outside reacts. The proteins in the membranes are rhodopsin in the rod cells, and opsins (photopsins) in the cone cells. There are three kinds of opsins—long, medium, and short—with two versions of the short opsin present in most human retinas. The rhodopsin in the rod cells is very light sensitive, and it gives us night vision while the opsins give us color. Each opsin pigment protein reacts with the specific waves of light that are reflected from the object being viewed. By "react" we mean that the light wave actually causes chemical reactions in the cone cells, and these chemical reactions are converted to neural messages that go to your brain. A cone cell with long wave opsins will react optimally when light waves of 564 nm hit them, and at lower intensity to light between 400 and 680 nm. Medium opsins react optimally with light of 534 nm, but lesser intensity around this peak, with a range of 400 to 650 nm. The short wavelength opsins react optimally at 420 nm, with a range of 360 to 540 nm. The overlap of wavelengths covered by these opsins ensures that information from light in the range of 360 to 680 nm is detected by the retina and thence transmitted, via the optic nerve, to the visual cortex of the brain.

The molecular structure of our Laphroaig bottle absorbs all light in the red and blue range, and some of it at the shorter end of the green range. Some of the light goes through the bottle, and some is reflected from the bottle and back to our eyes. Consequently, the green opsin proteins in our retinas are sending strong signals to the brain, which is what ultimately tells us that the bottle is a translucent green, while its label is a drab green. In other words, the brain is processing reflected light waves in the green range, and adjusting its interpretation of the color (in this case, to a dark green) from information that hits the short and long opsins. All of this information is put together in the brain's optical cortex to tell you what you are seeing. As this is going on, a lot of other neural processing is accomplished to make additional perceptual inferences about depth, lighting, shadows, and other aspects of vision.

You reach out to grab the bottle, and as you wrap your hand around it another surge of information starts to flood your brain. Your hands are amazingly well connected to its sensory cortex and in fact, if you drew a human figure with the parts of the body proportional to how much of the brain's sensory cortex is dedicated to that body part, the hands would be enormous. (Such drawings also show the lips, tongue, and genitals as much larger than the rest of the body.) Your hands are populated by a menagerie of specialized cells that detect specific kinds of tactile information. These cells are named for neurobiology pioneers such as

Merkel, Pacini, Meissner, Ruffini, and Ranvier, and each specializes in a different kind of touch. As your fingers reach the bottle, the Merkel-Ranvier cells in your fingers and hand are slightly displaced by the initial pressure of touching the bottle. This physical displacement causes a reaction in both the Merkel-Ranvier and Meissner cells, which are responsible for light touch and that tell your brain you have come into gentle contact with something. Next, as you grasp the bottle and your skin stretches to conform to the bottle's shape, your Ruffini cells respond, to tell your brain that you are putting more pressure on the bottle, information that is also being received from the Pacinian cells in the deeper parts of the skin layer of your hand. At a certain point the pressure on the bottle is strong enough for your brain to calculate that it is okay to lift it. Simultaneously, the thermoreceptor cells in your hand communicate to your brain that the bottle will not freeze or burn you. And all along, proprioceptor cells are telling the muscles of your arms, hands, and fingers where they are in space, enabling smooth movement. The same process repeats in your other hand, as you reach for the cap to remove it. From memory your brain knows to twist, and out comes the cork.

Laphroaig has a strong peaty odor. As the bottle is opened, a rush of molecules flows from its neck and out into the air. Some of these floating molecules reach your nose and are sucked inside as you breathe. Your nasal passage is lined with small olfactory "hair cells" that have two neat qualities. First, they are both close and directly connected to the olfactory bulb of your brain. And second, they have specialized molecules embedded in their membranes that act like lock-and-key mechanisms, capturing specific odor molecules floating out of the bottle. These membrane molecules, called olfactory receptors, act much as the opsins do in the retina of the eye. An odorant comes into contact with the receptor, and if it fits into the receptor protein a cascade of reactions will be triggered and a signal sent to the brain. That signal is then decoded by the brain into an olfactory experience. Each aroma or smell you sense, then, results from the interaction of many odorants hitting the olfactory cells of the nasal passage, and their subsequent interpretation by your brain. And more than likely, even as you sniff the Scotch in your glass, the parts of your brain responsible for memory are trying to break in with information about the odor. Marcel Proust wrote beautifully of the memories induced by smelling madeleine cakes fresh from the oven. Never known for his brevity, Proust indulged his readers with more than two thousand words on the experience in his *Remembrance of Things Past.* And in the context of ethanol, let's not forget that the aroma of this molecule may have been a valuable cue to our very remote ancestors.

Odorant molecules are incredibly variable, so our olfactory receptor repertoire is rather large. In humans, these receptors are coded for by some eight hundred such genes, while mice have almost twice as many. Each gene specifies a different sequence of the basic building blocks of proteins that are called amino acids. These different sequences affect the three-dimensional structures of the olfactory receptors and change the regions of the protein that will recognize an odorant. There is some controversy as to whether the reaction with odorant molecules involves physical contact or changes in vibration of the receptor molecules, but whichever mechanism applies, the structure of the receptor protein is intimately involved in odor reception.

The odorant molecules that are associated with spirits form a topic of growing research among brewers, wine-makers, and distillers, research that has been enhanced by gas chromatography—olfactometry (GC-O) or the E(lectronic)-Nose—a technique that allows researchers to break down an odor into its constituent molecules. It turns out that there are twenty key odorants in the nose of a peated single malt spirit, and nearly half of them are phenolics (molecules with structural rings of benzene with attached hydroxyl groups, or -OH). The structures of many such odorants are known, and they are distinctive enough that our very complex olfactory systems can probably detect all of them and transmit this information to our brains. This means that our perception of the odor that we associate with a peaty Scotch is based on the integration in our brains of twenty odorants. Beata Plutowska and Waldemar Wardencki, the scientists who discovered this, then went on to see how the peaty Scotch compared to other whiskies. By characterizing single malts, blended whiskies, and American whiskeys, they were able to show that each has a distinctive GC-O profile. Their final experiment used thirty-six different brands of whiskies from the four categories they recognized, and they concluded that sixty common volatile molecules offer everything needed to discriminate among the four types.

The smell of this peaty whisky alone is enough to get you to pour it into a glass (after all, we are descended from an ancestor that was attuned to aromas associated with alcohol), but the look of the beverage pulls you in further. Your retina is flooded with light reflecting from the beverage, as well as the glass. You use the touch cells in your hands to guide you to a firm grasp of the glass, and after lifting it with style, you propose a toast. "Clink!" The clinking glasses create a displacement of air that is proportional to the strength of the contact, sending air out in all directions like a wave. These are not the same kind of waves that produce colors, but they behave a lot like them. The clink's wave will have both a frequency (pitch) and an intensity (loudness). The wave frequency is measured in hertz units, with a hertz (Hz) defined as one wave cycle per second. Sounds with very high pitch are made of many wave cycles per second. Normal speech is at about 500 Hz; Mariah Carey can hit 3,100 Hz, and we reckon that the toast clink is at about 1,000 Hz (the same as a small bell ringing). By contrast, the clink's intensity is measured in decibels (dB), an electrical measure. Zero decibels are imperceptible, while about 120 decibels will cause pain. The toast clink is mellow, falling into a decibel range below 10 dB.

We hear through our ears, of course, and the way our ears achieve this feat is fascinating. Our hearing apparatus is truly a complicated gadget; it has several main parts, of which the cochlea is the most prominent. This organ is a coiled tube filled with fluid, and lined with small "hair cells." Connected to the cochlea is the oval membrane, which is in turn connected to a strange device made of three separate tiny bones called the stapes (stirrup), incus (anvil), and malleus (hammer), in order of their increasing distance from the oval membrane. These three tiny bones are also found in other vertebrates such as fish, but except in mammals the bones make up part of the jaw, meaning that our hearing apparatus evolved through the modification of bones that were originally in the jaw. Through this chain of tiny bones, the oval membrane is connected to a small tympanic membrane on the inside of the external ear canal. This picks up vibrations from the environment, which are then transmitted via those bones to the cochlea, where the hair cells convert them to electrical signals that go to the auditory centers of the brain for interpretation.

Ad hoc as it might seem, this evolutionary gadget works pretty well. The vibrations of the oval membrane travel through the fluid inside the coil, bending the hairs and generating a cascade of electrical signals that go to the brain. Since the cochlea is coiled, part of it is close to the origin of the vibrations, and part is farther away. High-pitched sounds are perceived only by the hairs at the beginning of the cochlea, while lower-pitch sounds are received all along the passageway. This means that the hairs at the beginning of the cochlea are subject to more wear and tear, leading to our tendency to lose our ability to hear the higher frequencies as we age. But don't worry; the clink of the toast isn't too high-pitched, and the sound of a great Scotch being shared will probably survive the ravages of time.

After the toast comes the *pièce de résistance:* taste. Our taste organs are commonly known as taste buds, and they are packed with little structures called papillae. Between two thousand and eight thousand of them are found on your tongue, on the roof of your mouth, on your inner cheek, on your epiglottis, and even on the upper part of your esophagus. The taste buds are collections of cells that are connected to one of the three nerves that carry information from the oral cavity to the brain. The sensitivity of the tasting apparatus varies among people, and there is some evidence that how many papillae you have is directly related to how well you taste. Those with exquisitely sensitive taste are called supertasters, while those who can barely taste at all are called hypotasters (or nontasters). But most of us are just normal tasters. You can get a quick idea of your tasting status with a simple do-it-at-home test involving counting the number of papillae on your tongue. Those with more than thirty papillae buds in an area equivalent to a single punch-hole made in a piece of paper are more than likely supertasters. A more accurate test of supertasting involves peoples' reaction to 6-n-propylthiouracil (PROP), a substance that is annoyingly bitter to a supertaster. Nontasters are oblivious to PROP, while to a normal taster the bitterness of PROP is detectable but not annoying.

It might sound as if being a supertaster would be a lot of fun, and for some people it is. For most, however, it means that they are often inconveniently hypersensitive to bitter flavors. Supertasters taste so intensely that many of them are picky eaters; and for that same reason many don't drink alcohol (because, as we will soon see, alcohol has a distinct influence on the taste buds).

There are five basic tastes—salty, sweet, sour, bitter, and umami—and researchers have discovered how all five work. Like smell, taste is chemosensory, meaning that the chemicals in the thing being tasted are detected by a receptor system. Each of the five major tastes has a different receptor. Salty, for instance, involves the detection of tiny molecules with charged ions. That's a roundabout way of saying NaCl or KCl. The best-known salt receptor is called ENaC (epithelial sodium [Na+] channel), but a second kind has also recently been discovered.

The cells in our body try to keep a balance of ions on the inside and outside

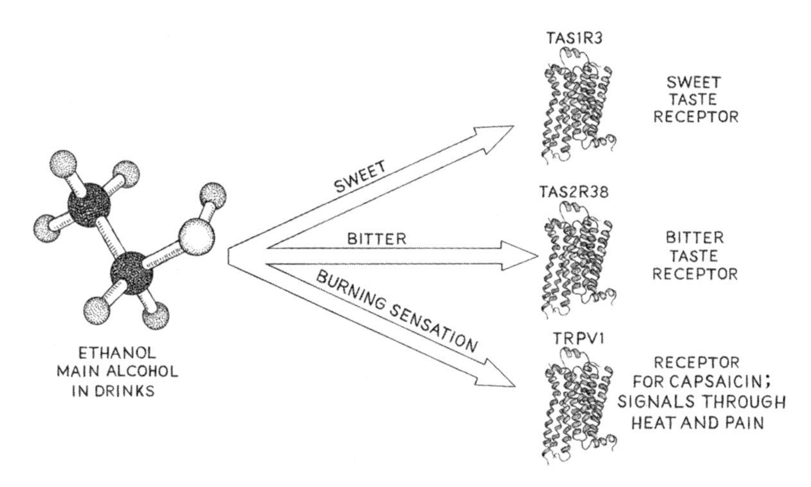

Figure 8.1. How we taste.

of their membranes. Consequently, molecular passageways exist in cell membranes that move the ions both ways, and these are called ion channels. The ion channel for salt deals with the Na+ ion of NaCl. The chloride ion (Cl^-) is not involved in salty taste reception via ENaC. For salty tastes, the taste bud cell is triggered by an excess of salt. The ENaC channel starts to move Na+ ions into the cell, and this starts a cascade of reactions that produce a nerve impulse that is sent to the brain, warning it of the presence of salt. Sour taste is also implemented by ion channels.

The other three tastes—sweet, bitter, and umami—use receptor systems that are similar to smell's. These tastes are detected by proteins embedded in the membranes of the taste receptor cells in the taste-bud papillae. Each of the three tastes has its own family of receptor proteins that are coded for by our genomes. For sweet there are nearly sixteen of these receptor proteins, and for bitter there are twelve. Umami has a set of three receptor proteins that detect small molecules like monosodium glutamate. Of the five basic tastes, sweetness and bitterness are the major players involved as the Laphroaig slides past the taster's tongue and palate (Figure 8.1).

One of the reasons it has proven so difficult to find substitutes for alcohol is that, contrary to popular belief, this molecule has a huge impact on taste perception. We guess the popular fallacy that alcohol itself lacks taste exists because a strong spirit like vodka is said to be tasteless. But ethanol actually has thresholds with respect to taste effect. Below a concentration of about 1.4 percent ethanol, the average person cannot taste it. In the range of 2 to 20 percent, ethanol will have elements of both bitter and sweet tastes. And over that threshold of 20 percent or so, a burning sensation comes along with those tastes.

The smokiness of Laphroaig is this whisky's overwhelming characteristic, and as you raise the glass to your lips, that smokiness comes wafting to your nose. This odor is important for your taste experience, because how we perceive taste is hugely influenced by our sense of smell. Indeed, since we have no taste receptors for smokiness, we don't really experience this aspect of our Laphroaig with our taste buds. Rather, it is our olfactory system that detects it, via the presence of guaiacol, a small phenolic compound that is emitted when things like creosote or peat are burned. When peat is used to smoke the grain during the malting process, some of this molecule makes its way into the mix and stays there because it is not removed during distillation. This tiny molecule is recognized by a specific olfactory receptor protein in the nasal passage, and it gets processed along with the taste sensation of bitterness and sweetness.

How about the burning sensation that alcohol imparts? Above 20 percent alcohol, it is caused by the ethanol overstimulating what are called "nociception" or pain receptors. For some reason, the same receptor that responds to capsaicin reacts to ethanol. A rather big molecule compared to ethanol, capsaicin is produced by most members of the plant genus *Capsicum,* the peppers, and reacts with a vanilloid receptor that signals pain to the nervous system. Because of this pathway, the burning aspect of the Scotch is not strictly a taste, but rather a sensation of heat-induced pain. But what about the distinctive juniper tastes of gins; the sweet, oaky tastes of bourbons; or the various tastes of infused vodkas? These highly characteristic tastes all overlay the basic bitter, sweet, and burning sensations of ethanol. Like the guaiacol in our peaty Laphroaig, each of these spirits has odorant and taste molecules that are characteristic of their category. These molecules are what stimulate our senses to make them appealing to us. All of the sensations perceived during our sip of Laphroaig are processed in different areas of the brain, which then puts everything together into a rich experience that interacts with our emotions, memories, and overall experience of life.

Let's return to that ethanol molecule we followed through the distillation process. When we last noticed it, in Chapter 5, it was sitting in a bottle of grappa.

As it turns out, its identical cousin was in our bottle of Laphroaig. When a glass of this Scotch was poured, that molecule of ethanol went into it, and it was among the millions that made it to our lips on the first sip. Avoiding becoming stuck to any of the receptors for taste or odor, it reached the esophagus, through which it also passed unscathed. And while many of its companion ethanol molecules were reacting with taste and vanillin receptors to give that bitter burning feeling, our particular ethanol molecule reached the stomach. There it met with a slew of the digestive enzymes that our stomachs secrete to break down our food for nourishment. One of the enzymes found in the stomach lining was potentially very dangerous for our ethanol molecule: the alcohol dehydrogenase (ADH) that we met in Chapter 1. There are actually five classes of ADHs, coded for by seven different genes in the human genome. Those our ethanol molecule needed to avoid were ADH1 and ADH2, which gobbled up and detoxified a slew of its companions. Our molecule luckily escaped the stomach lining and got into the bloodstream, via which it and the other surviving ethanol molecules could travel to the rest of the body.

Many ethanol molecules are transported to the liver, where there are more ADHs and more detoxification takes place, and where the compounds are metabolized into energy. An excess of ethanol over time is not good for any of your internal organs, and this is especially true for the liver, where prolonged abuse of alcohol will produce what is known as cirrhosis. This terrible disorder is essentially a scarring of the liver, since liver cells and tissues tend to repair themselves with nonfunctional scar tissues. These will accumulate as long as alcohol abuse occurs, until the entire organ ceases to function. Your liver, unfortunately, is not like your appendix or your gall bladder, which can be removed without causing too many problems if they malfunction. You really need a working liver. If caught early, cirrhosis can be brought under control—but only if alcohol intake ceases.

Our ethanol molecule, like many others, is still coursing through the bloodstream at this point, and might still wind up at any of many different destinations. One of them is the inner ear. The organs in the inner ear are responsible not just for hearing, but for balance as well. Tiny blood vessels can deliver ethanol to the little structures known as the semicircular canals. There are three of these canals: one for left to right balance, one for front to back balance, and one for rotational balance. The canals are filled with fluid, and they work mechanically in the same way the cochlea does, namely through the mechanical bending of hairs. When ethanol gets to these canals, it causes a swelling of those hairs that in turn causes dizziness, loss of balance, and the dreaded "spinning room" effect.

But our ethanol molecule makes it to the brain, settling in the cortex. This part of the brain is responsible for decision making and other important aspects of behavior. The molecule might equally have traveled to the amygdala (involved in emotional responses), the hippocampus (involved in memory), the cerebellum (involved in muscular movement and other important functions), or the striatum (involved in what are called executive functions), to name just a few regions of the brain. Wherever they find themselves in the brain, some ethanol molecules will reach the spaces between nerve cells known as synapses. The nerve synapse is essentially the conduit of communication between adjacent nerve cells. For our brains and nervous system to work properly, electric potentials need to be transferred among the neural cells that compose them. The synapses use molecules in the neural cell membranes to facilitate the transfer of neural information. Many small molecules such as Ca2+, K+, and Na+ facilitate this transfer, via channels in the membrane that are like the ion channels we mentioned in connection with the sense of taste. The receptors control the passage of ions into and out of neural cells, and thereby control the transfer of information across the synapse.

Ethanol is a versatile molecule in the brain because it can affect at least ten different kinds of neuroreceptors. These receptors control communication among six different kinds of neural cells, and hence influence a broad swath of the behaviors that are controlled by the brain. According to Karina Abrahao, Armando Salinas, and David Lovinger, ethanol has two major effects on the brain. Acutely, it is both a stimulant and a depressant. Chronically, it will cause changes in the brain and hence in behavior that may culminate in addiction (alcoholism). Our ethanol molecule will be directly responsible for some acute changes, and if abuse occurs it will also contribute to longer-term addiction.

Our ethanol molecule is one of many that make it to the synapses of the cortex. Once there, it interacts with a receptor protein called a GABA (gamma-aminobutyric acid) receptor. Normally this receptor binds the relatively small GABA molecule ($C_4H_9NO_2$), and this binding in turn regulates synaptic communication of neural cells. Our ethanol molecule disrupts this normal GABA interaction by increasing the passage of ions in the synapse where it settled. At the same time, all of the other ethanol molecules that made it into the bloodstream after that first drink are also increasing the activity of GABA receptors. This causes an overall increase in synaptic activity and an imbalance in the neural signal. For instance, alcohol can increase the effect of the GABA receptor, which inhibits neurotransmission, or it might reduce activity in the mGlu receptor, which usually enhances neural communication. In other words, alcohol can affect various ion channels differently, but one thing is for sure: the combination of effects on the synapses of the brain's cortex leads to impaired behavior. Similar issues arise in other regions of the brain where alcohol penetrates; if it goes to the cerebellum, for example, muscular coordination is impaired, while ethanol in the amygdala causes an altered emotional response.

Alcohol has a further effect on our brains: it stimulates the neurotransmitters dopamine and norepinephrine. Alcohol amps up the production of norepinephrine, which results in heightened arousal and subdued inhibition—but only in the early stages of drinking. Dopamine is also stimulated by ethanol, and more dopamine means more activity in what are known as the reward circuits, leading the drinker to seek more ethanol. Dopamine is also involved in memory, so alcohol's propensity to increase dopamine means that the memory circuits of the brain further seek the buzz of ethanol on the brain after that first drink. Ethanol's impact here should be obvious. It increases our impulsivity and depresses our inhibitions, muscular activity, memory, and emotional response; and, at the same time, we crave more of it. If unchecked, this acute craving can lead to long-term chronic effects, and in some cases, alcoholism.

All of these various effects of alcohol on the brain are mediated by the concentration of ethanol in our bloodstreams, and for the sake of explanation we have focused on what happens as alcohol is imbibed in excess. In lower concentrations, the effects of alcohol are both less extreme and more benevolent than those we have described; in fact, the same mechanisms of brain biology that cause problems when we drink too much are responsible for the pleasurable relaxation and mild disinhibition that the more judicious intake of alcohol can provide. Add to this that alcohol is the most effective vehicle there is for a whole variety of agreeable sensory experiences, especially in the realms of olfaction and taste, and it is clear that our human experience would be much the poorer without this molecule. (To get a sense for just how much poorer, we need look no further than the sneaky, passionate excesses of Prohibition.) Having written three books on alcoholic beverages, we clearly believe in these effects as significant enhancers of human existence. The dangers of excess are undeniably out there, but with carefully monitored moderation, you can enjoy that alcohol buzz without falling into the trap that our ethanol molecule and its billions of fellows have laid for you.

MARKET LEADERS

The "Big Six"

9

Brandy

Alex de Voogt

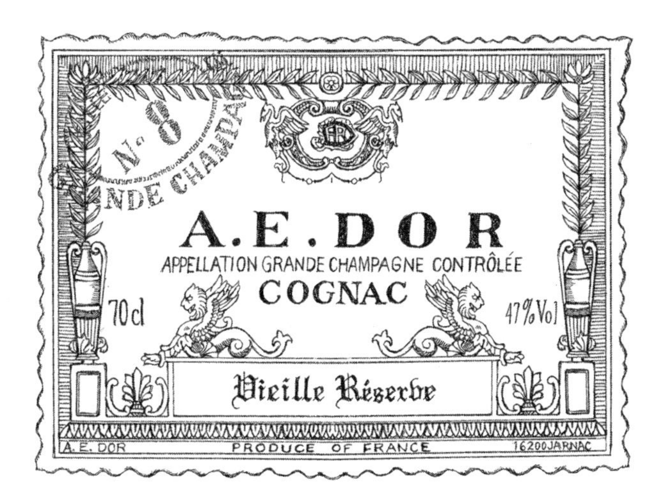

The red number 8 stamped on the label was our first clue to the quality of the Cognac in the tantalizing bottle in front of us. It referred to the Hors d'Age range, the lowest being 6 and the highest 11, that indicates how far above the usual Cognac-quality scale this one lay. Think of it this way: the house of A.E. Dor has been blending Cognacs since the nineteenth century, and the base wine for the Hors d'Age range comes only from Grande Champagne, the most revered Cognac wine region. The midrange number 8, the one we're looking at now, has a blend of Cognacs aged forty years and longer. With ethanol making up 47 percent of each sip, this is the strongest Cognac the house offers. Yet although the higher alcohol percentage removes much of the smooth sweetness that Cognacs can possess, this spirit is not aggressive, but instead is spicy and full of flavor. Take a small sip first to adjust to the alcohol, then a bigger one for an immersion in flavor—now wait several moments for the accents to change during the long finish. I have long favored this Cognac over the many others I've tried, both old and young. Savoring a flight of Cognacs not only lets you directly compare strengths of flavors and lengths of finish; it can also help you find your own favorite blend.

In the southwest of France, the medieval town of Cognac has bestowed its name on a brandy that is both eminent and luxurious. Grand houses with names such as Rémy Martin, Otard, Camus, Hennessy, and Martell line the streets of the ancient city center, offering visitors a glimpse of their renowned distilleries. A two- to three-hour stroll to the east and you will find, in the town of Jarnac, another group of Cognac houses with names such as Courvoisier, Hine, and A.E. Dor.

Late one morning, a friend and I arrived at A.E. Dor unannounced, and were greeted by an elegant woman who occupied an unassuming office. She called her colleague, an elderly and cheerful gentleman, to welcome their two Dutch guests. He took us into a small, charming tasting room consisting of a counter with a long row of Cognac bottles just behind. There was no distillery to visit, we were told, but perhaps we were interested to see the *paradis* before we tasted some of their lofty spirits? Certainly we were! We followed the man through a gate locked with a chain and proceeded beyond, to a building that seemed little more than a large barn. Inside, on low wooden platforms under a beamed roof, was a large collection of glass demijohns with wicker covers. These Cognacs had not moved since their collector, Amédée-Edouard Dor, had set them there. In 1858, Jean Baptiste Dor founded a modest Cognac house, and it was his eldest son, Amédée-Edouard, who would add his initials to the label in 1889. His passion for collecting and selecting the most esteemed Cognacs of his time would give this house its reputation. For decades he used oak barrels to age rare samples, mostly from Grande Champagne, today the most esteemed part of the Cognac vineyards. He then decanted the results into the glass demijohns in which their essence would be preserved for history.

We walked past the first glass containers, each of which was marked with year, name, and region, such as Fine Champagne or Grande Champagne. The farther in we strolled, the further back these Cognacs went in time. Only some had names; and when we lingered in front of an 1834 Prince Albert, we still had a way to go. We finally stopped in front of a jar with a sign that said: "Grande Champagne, 1811, 31% volume." We had reached Napoleonic times. Next to it, a slightly smaller jar placed on a wooden box said "Grande Champagne, 1805." We were back to the start of the Napoleonic Wars, and the finish of our tour in paradise.

When we returned to the counter for a tasting, a VS (very special) was served in small tasting glasses with golden letters spelling "A.E. Dor." The master blender of a Cognac house selects different ages and different crus, or subregions of the Cognac vineyards, to create a character that fits the house. The youngest or least expensive of a range is often the least balanced, but at A.E. Dor the young ones already have great character. In the tradition of Cognac, the range continues with VSOP (very special old pale) and XO (extra old), requiring a minimum of four or six years of aging, respectively. In between there was also a Napoleon, a most appropriate epithet after a visit to the *paradis.* Several others were passed over as our host generously insisted on serving at least a sip of each of the house's top Cognacs, numbered from six to eleven. Just the nose of number six immediately transported us to a different world, and introduced a complexity absent in the previous tastes. We progressed to other bottles in this range; and number eight, partially due to its 47 percent alcohol instead of the 40 percent found earlier, added yet another layer of pleasure. Despite the expense of these bottles, our enthusiasm during this free tasting made the host offer us also a sip of number eleven. Here, sheer impact gave way to subtlety. We left with a historic intoxication on our palates, a selection of two bottles each, and, as a gift, two tasting glasses.

In May 2020, one of three remaining bottles of a 1762 Gautier Cognac, described as the oldest surviving Cognac in the world, was sold at auction for a record sum of $144,525. With the bottle and label still in good condition, this Cognac, made before the French Revolution but using a production process used today, was still good to drink. The Cognac industry as we know it today is not much more than a century older than that venerable bottle, having been spurred into existence by Dutch traders. As summarized by L. M. Cullen, "The great consumers, importers and makers of spirits in the seventeenth century were the Dutch, the richest people of their day" (for more on those early days, see Chapter 3). They had the most advanced distilling industry of the time, one devoted to converting a cheap wine surplus to profitable brandy; and they not only made spirits in large quantities in Holland, but were also the main buyers elsewhere. In France, the Dutch often accounted for more than half of the brandy exports during the seventeenth century, often sending it as far away as the Baltics as well as supplying a growing home market. Like other seafarers, they found that adding brandy to a wine barrel prevented it from turning into vinegar on long voyages, while creating an excellent tipple. It was Dutch demand that lay behind brandy distilling industries all the way to the Mediterranean, but the traders did not themselves settle in those faraway places. It was their home consumption that made the industry flourish and specialize, so that the high-end Cognacs and Armagnacs took a place in the Netherlands market next to pomace and fruit brandies from Italy and Germany, rums from the colonies, and their own Dutch genevers.

The British, unlike the Dutch, were mostly at war with France during the seventeenth and eighteenth centuries, so they mainly concentrated on distilling at home. Even so, early in the eighteenth century, John Martell from the British island of Jersey started his famous Cognac house, which continues today. His earliest labels identified his blends as "Old Brandy," "J. & F. Martell," and "Cognac," and though most of his exports went to England, some also found their way to the fledgling United States soon after its independence. Frédéric Martell, John's grandson, is credited with the first use of the VSOP designation, which was applied to a shipment to England. Martell Cognac was served at English coronations and at many other great occasions, further sealing a strong link between Cognac and the Anglo-Saxon elite. In 1765 Richard Hennessy, an Irish officer who had served in the army of Louis XV, founded the Cognac trading house that bears his name, and by 1794 he too was shipping his product to the Americas and receiving a royal blessing: in 1817 the Prince of Wales, later King George IV, requested the creation of a Hennessy VSOP Cognac. With its longtime association with royalty, statecraft, and the elite, Cognac has secured its reputation as a sophisticated and high-quality spirit.

The international popularity of Cognac spurred the production of other grape brandies around the world. These spirits cannot claim the title of Cognac, but they emulate its prestige. Perhaps the most unusual are the products of the

Yerevan Brandy Company, which until 1940 was known as the Shustov factory, but now, like Martell, is owned by the French conglomerate Pernod Ricard. These brandies, sold under the ArArAt brand since 1887, are made in Armenia from grapes grown within sight of the eponymous mountain. In the early twentieth century the company became the main supplier of the Russian imperial court, and in the 1950s it produced "Armenian Cognac" for all regions of the Soviet Union. This Armenian brandy, which is today still referred to as "Cognac" by many of its clients, became one of Armenia's main export products, and is sold by brands such as Akhtamar, Ani, and Nairi. Its association with the Russian elite led to a level of fame and success for the Armenian brandy in the former Soviet Union that rivaled that of Cognac in Western Europe.

Every major Cognac house prides itself on its paradis, which shows off the depth and breadth of its blends. The name of the house is mostly associated with the work of a master blender. In the case of Hennessy, a family in the line of Jean Fillioux oversaw the aging and blending of Cognacs for eight generations. While the blender is all-important to a house, the vineyards from which the base wines come are not always closely linked. Even the very large brands, such as Hennessy, Rémy Martin, and Martell, do not own nearly enough vineyards to support their vast production. Most firms do not buy grapes to fill the gap, but instead purchase the wine from which the Cognac is made or, like Courvoisier and Hine, acquire young Cognacs shortly after distillation, and blend them. So although the growers and nowadays the winemakers are closely monitored by the grand houses, there is hardly any vertical integration of the production process.

Cognacs and Armagnacs, unlike most other spirits, are distilled from white wines. Other grape brandies such as marcs and grappas are commonly distilled from the residues of wine pressing, including the grape skins, pips, and stems that often impart a harsh taste. All grape brandies are considered eaux-de-vie in a general sense; but the Dutch *brandewyn* (burnt wine), from which the term brandy is derived, precisely captures the production processes of Cognac and Armagnac, because they are distilled from wine, and from wine only.

In the past few decades, the wine-making part of the Cognac production process has been given increasing attention, further improving the refinement and complexity of the final product. Cognac may be made only in the Cognac region, with strictly defined sub-appellations such as the prestigious Grande Champagne, Petite Champagne, Borderies, Fins Bois, and the less esteemed Bois Ordinaires and Bois à Terroirs. Its Gascon cousin Armagnac is similarly confined, with three appellations: the traditional Bas-Armagnac, Ténarèze, and the smaller Haut-Armagnac. Their unique environmental contexts, or terroirs, determine the quality of the wines from which the brandies are distilled.

Originally the Balzac and Colombard grapes predominated in Cognac, but they were largely ousted in the nineteenth century by the more acidic Folle Blanche, a variety that produced a highly aromatic and fragrant spirit. In the 1870s the vineyards were stricken by the infamous phylloxera louse, and while Folle Blanche tops were grafted onto American rootstocks in response, the result proved susceptible to gray rot. Hybrids were also tried, but they produced unpleasant aromas. Consequently, while the traditional grape varieties are still grown and used, most Cognac is now made from the originally Italian Trebbiano grape, better known in France as the Ugni Blanc. Since Cognac is far to the north of Italy and is, indeed, the farthest north this grape variety is grown, the grapes remain green and acidic compared to their Italian counterparts. Even after experiencing the long but temperate summers of the Cognac region, they are still high in acid and not completely ripe when they have to be picked in late October, before the risk of severe frost becomes too great. The grapes are gently harvested by machine, using vibration so that the branches do not get harvested as well, from vines that are ideally aged between twenty and thirty years, as opposed to the sixty years one would find for classier wines. In the Armagnac region to the south, the hybrid Baco grape proved successful in the recovery from the phylloxera episode, and it is now used in addition to the Colombard and some Folle Blanche and Ugni Blanc. Various blends of these varieties go into the wines destined for distilling, so that Armagnacs tend to differ more in aroma and intensity than Cognacs do.

Although the vinification of the initial wine has recently grown in complexity, historically the wine destined to become Cognac had just over 8 percent alcohol. Distillation increased alcohol concentration a maximum of nine times, to no more than a prescribed maximum of 72 percent, from which the alcohol concentration was then reduced through barrel aging. In contrast, Armagnac is initially distilled to a lower strength of 52 percent alcohol. The base wines for both are very acidic, and in Cognac they typically undergo a malolactic fermentation to convert the harsher malic acid to the softer lactic acid before distillation. Distillation usually commences six weeks or less after fermentation is completed.

By the end of the seventeenth century, with help from the Dutch, Cognac producers had figured out how to double-distill in pot stills to eliminate most of the impurities, a process which involved having the spirit pass through a "half-strength" stage. (Armagnac, in contrast, is typically made without this intermediate stage.) And to accommodate the needs of small farmers, Cognac stills used to be of limited size. But that has now changed: the initial still can hold 140 hectoliters, though maximally filled with only 120. And if the spirit is to be called a Cognac, the second distillation is legally limited to 25 hectoliters, fitted into a 30 hectoliter still. This starting amount is reduced to around seven hectoliters of Cognac in the second pass, although another six hectoliters or so can be redistilled. Even though the half-product is usually not tasted for quality, the principal chemical reactions that determine the quality of the Cognac are thought to occur during the first distillation, with their products simply concentrated in the second stage. Variations in this basic distillation process are endless, and mastering them is considered an art. Also helping to determine the character of any Cognac are many other choices, traditions, and legal requirements. For example, by law, stills may be heated only from outside, requiring massive brick ovens with an open fire; and the resulting complex and curvaceous "Charentais-style" copper still, unique to Cognac production, has remained largely unchanged from the beginning.

In contrast to Cognac, the "alembic ambulant" was often found in Armagnac. This was a transportable still that could be wheeled into a barn for use by farmers who could not afford their own, and it provided a single continuous distillation process. Double distillation is also sometimes used in Armagnac country, but most prefer the aromatic power of single distillation. Some argue that while Armagnac may be rougher when young, it ages better than Cognac. Ultimately, it all depends on the skill of the distiller. In Armagnac those distillers have a choice of stills, and experiment accordingly. When making blends, they wait a full five years before assembling their product.

The longest stage of the production process begins when the distillate is finally put in French oak casks that instill both color and a taste (see Chapter 6).

For Cognac, barrels must come from the forests of Tronçais or Limousin in central France, whereas makers of Armagnac often source the white oak locally. The barrel-makers are an independent industry and Vicard, for instance, not only produces and sells new French oak casks and barrels, but also provides used Cognac barrels to anyone interested. I myself have tried aging a Dutch genever (made in The Hague by Van Kleef) in used Cognac barrels: the experiment turned out beautifully, with a deep color and refinement of the genever taste. High-quality Armagnacs are often sold indicating a vintage year, but Hine is one of only a few Cognac houses that has regularly offered bottles with a vintage year. The 1953 Hine is heralded as one of the best, with universal praise for the 1964, 1975, and 1988 as well; 2002 is a more recent but particularly good vintage. In the A. E. Dor paradis each demijohn is given a vintage year, because no blending took place after its meticulous aging by A. E. Dor himself. But although it is generally believed that any year can produce a great Cognac (what differs among years is the amount of excellence that can be produced), for Cognac blending is the rule.

The method used in barrel aging distinguishes several brandies, including the Spanish "Brandy de Jerez" and "Brandy de Torres." By law, Brandy de Jerez needs to be matured for at least three years in barrels that previously held sherry. Some casks that are over half a century old are still in active use. Spanish brandy-makers are also known for using the solera system, commonly used to age fortified wine. In this method, a quarter of the contents of the oldest barrel is removed and bottled, to be replaced by the next oldest, and so on until the youngest barrel is replenished with newly distilled brandy. It requires a large set of barrels to provide this continuous, distinctive, and consistent production. A Spanish brandy labeled twenty years old might have a blend including four- to fifty-year-old brandies, with twenty being the weighted average. In Cognac, any age given would indicate the youngest component. And while the master blenders of Cognac apply their skills after maturation is complete, the Spanish blend during the aging process.

Spanish brandy is mostly made around Jerez in Andalucía, but some comes from Penedès in Catalonia; the two regions have rather distinct climates. In Andalucía they use the Airén and Palomino grapes, the latter also common in sherry production, while in Penedès they use grapes such as Macabeo (also used to make sparkling wines) in addition to the Ugni Blanc also used in Cognac and elsewhere. Andalucían brandy is mostly single-distilled, although the main producers in Penedès double-distill as well. Together with Cognac and Armagnac, Brandy de Jerez is one of only three regulated brandies in Europe, and it may be blended and aged only in the region known as the "sherry triangle" that lies between the cities of Jerez de la Frontera, Sanlúcar de Barrameda, and El Puerto de Santa María.

South African brandy boasts a history almost as long as that of Cognac. The presence in South Africa of the Dutch, who had dominated the early trade in spirits and whose enthusiasm had supported the development of the high-quality Cognacs, was a big reason why. It is said that in 1672 a Dutch sailor produced the first *brandewyn* that got the industry going. The wine is different from that used in the Cognac region; higher sugar levels raise the alcohol in the base wine from 10 to 12 percent, in contrast to the 8 percent in Cognac, and are said to give the final product more fruity flavors compared to Cognacs. For their acidity, South Africans use the widely grown Chenin Blanc and Colombard grapes, another nod to the Cognac tradition that sometimes involves Colombard grapes. Stills for South African brandy are also bigger, using steam coils because there is no legal requirement to heat them from the outside. Both French and American oak are used, and South African brandies are double-distilled as in Cognac. Brands such as Oude Molen ("Old Mill" in Dutch) and Van Ryn, the top range of the Distell distillery, remind us of the spirit's Dutch heritage.

Producing brandy, or distilling wine, is likely to be done in any wine-producing country. In Peru and Chile, for example, there is pisco, a brandy distilled using local grape varieties and relatively rarely aged in oak. The distillation process was introduced by Spanish settlers to replace imports of liquor, and pisco is now claimed by both Chile and Peru as the national spirit. Greek brandy is often known as Metaxa for the brand created in 1888 by Spyros Metaxa, a well-traveled silk tradesman; technically it is only arguably a brandy, because it blends separately oak-aged Muscat wine with wine distillates and adds Mediterranean botanicals. Some other regions of the world imitate the process developed in Cognac, while others simply adopt a name resembling "Cognac." Most famously, "conhaque" in Brazil survived a lawsuit, and the name is allowed to continue in use for spirits as long as it does not refer to a distillate from wine. Other examples include Konyagi, a spirit produced in Kenya and Tanzania. This is distilled from molasses and, like conhaque, is hardly a drink for the elite.

Cognac is the main French alcoholic export product by value, more so than Bordeaux still wines and Champagnes. The French pomace brandy known as marc, in contrast, is widely made but mostly destined for local markets. In Italy the pomace brandy known as grappa traditionally cornered the local spirits market, but never even began to compete with Italy's wine exports. The absence of wine production in the Netherlands and the United Kingdom boosted grain-based domestic spirits such as whisky, gin, and genever instead, while in their tropical colonies both countries, like France and Spain, produced rum from sugar cane. Naturally enough, only colonial wine-producing areas such as South Africa, or Peru, were able to engage in grape brandy production.

Since the Dutch began expanding the market for brandies and other spirits from the sixteenth century onward, followed by the British in the centuries after, each spirit-producing region has developed a particular strength that has allowed it to survive in the international marketplace. Today, a whopping 97.7 percent of Cognac production is exported, with the United States by far the largest market (87 million bottles in 2018), followed, amazingly enough, by tiny Singapore (27 million bottles). In Hong Kong, Cognac has been the liquor of choice at wedding banquets since the 1970s and is a common luxury gift. The spirit's gifting role is what makes the city one of the top consumers of Cognac, even though, according to the anthropologist Josephine Smart, there is generally only light alcohol consumption among the local population. In short, the socioeconomic and sociopolitical shifts in the world, and the continuous association with (emerging) wealth and prestige, remain highly visible in the history of brandy in general and Cognac in particular.

In contrast to rum and gin, brandy is considered an after-dinner drink in most of Europe, although it is often drunk with the meal in Russia. The British make brandy butter, and still pour brandy over their Christmas puddings and set it alight, a tradition celebrated by Charles Dickens in *A Christmas Carol.* Americans, for their part, revere their brandy cocktails, including the Brandy Alexander, Brandy Daisy, Metropolitan, Vieux Carré, Sidecar, and Between the Sheets, some of which may also contain rum and whisky. The ways in which brandy is used in different cultures is perhaps as varied as the brandies themselves.

The gastronomy of cooking with Cognac, however, is steeped in a distinctly French tradition. The French dish known as coq-au-vin, or rooster in wine, illustrates at least four different uses of brandies (Cognac specifically) in cooking. A fowl made in this style may be marinated in wine to which Cognac is often added. Cognac may be also added when simmering the dish in the oven or on the stove. The cook may also wish to use Cognac to dissolve the food residue left in a skillet after the browning of the chicken, in the process known as deglazing. Finally, Cognac and other brandies are perhaps best known for the flambé process during which, after heating in a skillet to volatize the alcohol, they are set alight. A browned chicken in coq-au-vin style may be flambéed before being put in a pot to simmer. A Christmas cake can be set alight by pouring some heated Cognac over the top, while Crêpes Suzette and several other dishes may be flambéed with different kinds of eaux-de-vie and liqueurs, depending on one's taste.

The grand Cognac houses do not promote their noble product as an ingredient for a chicken dish, nor do they mention the medicinal merits of their lofty liquor for the common cold. Cognac-makers take themselves more seriously, as purveyors of a modern luxury item. In a recent advertisement by Rémy Martin, men and women of all ethnic backgrounds are shown in chic clothing while brandishing Cognac-filled snifters. Hennessy, too, features young, stylish women lounging in the backs of limousines, or exiting a luxury hotel. Back in 1968 it showed a couple preparing to play tennis, with the tag line "The taste of success." Recently, Courvoisier showed a portrait of Napoleon behind a table with a bottle of its elixir, a partly filled snifter, and a candle softly lighting the dark furniture in the room (Napoleon did indeed take several barrels of Cognac with him to Saint Helena). In other words, Cognac is international, and steeped in history and luxury. Its competitors emulate the Cognac advertisers, with ArArAt promoting slogans such as "Created by time, becomes history." In 1991, Brandy de Jerez boasted its hoped-for medieval heritage when, under an image of a sixteenth-century ship, it proclaimed: "In 1522, while Juan Sebastián de Elcano became the first man to sail around the world, the world's first brandies had happily been around for 600 years." Brandy's long history, its international allure since the beginning of the age of European exploration, and the pride of each country that produces it, all swirl in that bulbous short-stemmed glass, ready to intoxicate you with every sip.

10

Vodka

Ian Tattersall

It seemed pointless at first to be sampling varieties of vodka, a spirit prized for being neutral. But soon we found ourselves faced with a bewildering array of choices, each label touting unique virtues. Captivated by visions of ice-encased bottles being lovingly extracted from freezers, we eventually settled on a vodka from Iceland. The brainchild of a noted Scotch distilling house, this spirit is produced in a single pass from a wheat and barley mash and, at 40 percent ABV, is possibly topped off with a bit of neutral spirit. It uses pure glacial meltwater filtered through a porous lava field, in a process that employs geothermal energy to heat a three-thousand-liter copper Carter-head pot still much like those stills developed in the nineteenth century to rectify gins. Instead of containing botanicals, the still's bulbous head houses crushed lava; and, once condensed, the liquid passes through yet more lava before being bottled. With only a single distillation we hoped to taste the grain, and we were not disappointed. Absolutely clear, this vodka had slender legs that trickled slowly down the glass, and a touch of lavender and Alpine herbs on the nose. It tingled the inside of the lips, and hit the front of the palate unctuously, with elegant traces of barley malt and a hint of spice. Two drops of bitters and an ice cube added significant depth, transforming this gentle spirit into a complex drink we could happily have nursed for the entire evening.

Vodka has a bit of an identity issue, even if its elusive nature has proven a marketing boon. As defined by the Alcohol and Tobacco Tax and Trade Bureau of the U.S. Treasury, this popular spirit is "neutral spirits so distilled, or so treated after distillation with charcoal or other materials, as to be without distinctive character, aroma, taste, or color." In turn, "neutral spirits" are "distilled spirits produced from any material at or above 190° proof, and, if bottled, bottled at not less than 80° proof." In other words (leaving aside its hardly unique alcohol content), vodka is defined in the bureaucratic mind by what it isn't, rather than by what it is. Perhaps this is inevitable, because the innumerable spirits sold as vodka in countries around the world may be made from virtually anything that can be fermented, including grains of all kinds, molasses from sugarcane or beets, potatoes, and an array of fruits including apples and grapes. A couple of producers even use milk. What all those vodkas have in common is that they have passed through some kind of process—normally filtration, or multiple distillations, or both—that is designed to remove as many as possible of the impurities that are inevitably left after the initial pass. In theory, these various gambits should indeed leave behind a "neutral spirit" that lacks any gustatory qualities except the famously warming effects of alcohol on the tissues of the oral cavity. Yet in practice their efficiency is never 100 percent; and historically (even before the fad for fruit vodkas), some distillers and compounders have not been above exposing their vodkas to tiny quantities of honey, sugar, glycerin, citric acid, and other ingredients in order to influence their flavors or textures. Currently, such practices are forbidden by law only in Poland, and many would argue that there still is gustatory variation even among Polish vodkas.

However their chosen spirit achieves its desired distinctiveness—whether through ingenious distilling, or by dint of clever publicity—one of the most remarkable things about vodka drinkers is their tendency to be fiercely loyal to their preferred brands. It is inarguable that some vodkas, at least, vary subtly not only in taste but in mouthfeel as well. But at the same time, vodka's vaunted neutrality makes it very attractive to those many cocktail lovers who want their drinks to have a stiff backbone while fully expressing other flavors. In the relatively short time since vodka became an "international" spirit, this two-tiered appeal has led to huge marketing success: around 1976, vodka overtook whisky as America's best-selling spirit, and it has retained that distinction ever since.

This international celebrity for vodka is a long way from its humble beginnings as a regional spirit in Russia or Poland or thereabouts—a place of origin that will be endlessly debated, no less passionately than Georgians and Armenians argue about where wine began. The Russians probably have the edge in establishing the earliest documented date, at least if you take a really broad view of what vodka is. Russian peasants appear to have been producing a strong spirit as early as the ninth century, well before the introduction of the alembic, employing a variant of the "freeze-distilling" method that a handful of brewers use today to produce "beers" with an ABV as high as 67.5 percent. The procedure used in medieval Russia apparently involved fermenting mead in the fall, then leaving it outside in the winter, picking out the ice as it formed at the top of the open barrel. This technique progressively removed water from the liquid, increasing the alcohol level in what was left behind. Whether or not the doubtless rough result would truly qualify as vodka is of course debatable, but an 1174 document known as the Vyatka Chronicle is said to note the existence of a distillery in the remote Russian town of Khylnovsk, some five hundred miles to the east of Moscow. This was only a couple of decades after the very first cryptic Western mention of distilling in the Salerno *Mappa Clavicula,* and just how knowledge of distilling might have penetrated to the eastern frontier regions of Russia at this early date is unclear. It is conceivable that there was Chinese influence; more likely, the product involved was made using some refinement of earlier methods. By 1430, though, something much more closely resembling modern vodka does seem to have been produced by monks in the Chudov monastery that resides within the Kremlin walls in Moscow—although, sadly, a great deal of early documentation relating to the development of Russian vodka was destroyed in the seventeenth century when the Orthodox Church took action against this invention of the devil and tried to eliminate all mention of it from the record.

Wherever it may have originated, and however exactly it was made, a highly alcoholic Russian mead derivative clearly proved the perfect antidote to the country's notoriously harsh winters. So much so, indeed, that by the middle of the fourteenth century a British diplomat was moved to describe this beverage as Russia's national drink. The potion he remarked on still appears to have been subjected (perhaps inevitably) to winter freezing; and it is recorded that isinglass, a fish-bladder extract still employed by winemakers today, was used to clarify it. By 1505 the first distilled Russian vodka exports had been sent to Sweden, where the drink was already popular, having apparently long been consumed there for medicinal reasons. It was also fashionable as a tonic back home in Russia, where it was described in the 1533 Novgorod Chronicles as a restorative *zizhnennia voda* (water of life). In the mid-sixteenth century, by which time alembics were widely available, Russian apothecaries had begun to refer to the spirit as "vodka [little water] of bread wine," signifying a spirit made from grain rather than from mead. Eventually, a relatively standardized manufacturing procedure was developed at the top of the market that involved an initial double distillation of a wheat and/or rye mash, followed by the addition of milk and an intermediate distillation, then dilution with water and the addition of flavorings before a final pass through the still.

In 1474, Tsar Ivan III established the first Russian state vodka monopoly. This measure had fiscal motives, and created a dangerous government dependence on alcohol taxes. It also inaugurated an official Russian attitude toward vodka that has since vacillated between moral and pragmatic extremes. As the heavy hand of the aristocracy leaned ever harder on the peasants and serfs, the consumption of spirits helped to alleviate the rigors of their lives. But it also led to widespread alcoholism, debilitation, and misery among the very people whose labors made the rural economy function. Eventually, a significant portion of the populace also fell into debt to the government-run *kabaks* (taverns) that had been created in 1552 by Tsar Ivan IV. This Ivan (the infamously terrible one) had reformed the aristocracy, and the taverns were intended initially to reward his new supporters with exclusive access to vodka. But things eventually got out of hand, and by the time the seventeenth century came around there was a kabak in every village, supplying all classes and often on credit. This system lasted until a currency reform under Tsar Alexei in the mid-seventeenth century caused huge inflation, and the cost of a pail (12.3 liters!) of vodka rose from well under a ruble in 1653, to five rubles a decade later. Apart from inciting riots, this price increase also had the effect of driving distilling underground, into private homes and vodka dens. In these places it clandestinely flourished until 1716 when, over the objections of a Russian Orthodox Church that was doing its best to suppress the demon drink entirely, Peter the Great (himself a two-liter-a-day man who believed that wives should be whipped for dragging unwilling husbands out of taverns) decreed that vodka production should be the exclusive province of aristocrats, who paid heftily for the privilege of exercising that right.

Peter's efforts to limit them notwithstanding, distilleries eventually proliferated again, especially under Catherine the Great in the late eighteenth century. During her reign a huge technological advance also occurred, made necessary by a general decline in quality as the multiplying numbers of vodka producers cut corners to pay for their expensive licenses. Filtration of the usually pretty rough spirit had long been performed inefficiently through sand or layers of felt; but in 1780 the Saint Petersburg chemist Theodore Lowitz introduced filtration through hardwood charcoal, which created a much purer spirit. Legend also has it that, shortly before his death in 1894, Tsar Alexander III asked the distinguished chemist Dmitri Mendeleev (yes, he of the periodic table of the elements) to suggest how to improve the quality of the vodka available. Mendeleev is said to have responded with the memorable recommendation of dilution to 40 percent ABV, which remains the standard today. Alas, although Mendeleev's doctoral dissertation had indeed involved both alcohol and water, it appears this was a myth: something like the 40 percent standard had already been imposed by tsarist edict back in 1698. Still, however it may have been arrived at, the 40 percent solution proved to be a magical one, providing the ideal viscosity of the spirit both at room temperature and—especially—straight out of the freezer.

Meanwhile, vodka in Poland was pursuing an entirely independent course. For a start, unlike Russian vodka's origin in mead, Polish vodka's origin may have been in wine. From the eleventh century onward, and possibly earlier, Polish peasants had cheered themselves up with a product they called *gorzalka*. This was most likely a freeze-distilled wine (records of viticulture in Poland go back to the tenth century), though the term may also have covered some mead or grain products. Whatever the case, the name derives from the verb "to burn," suggesting that the stuff was alcoholic enough to catch fire when lit. The term *wodka* first appears in the written record in 1405, in court documents that refer to alcohol-based cosmetics and medicines. Evidently, the distilled wheat-based beverage that was undoubtedly already made at the time was known by *gorzalka* or some other term, and it was only later that the name *wodka* became associated with it.

Distillation technology was probably introduced to Poland by German Hanseatic traders around the end of the thirteenth century. By the early fifteenth century, distilling was apparently widespread in the country, and by 1534 the renowned herbalist Stefan Falimierz was loudly praising the ability of distilled spirits to arouse lust, and thus to promote fertility. But beyond this we don't know a lot. To give you some idea of how hazy the early history of Polish vodka remains, it is widely and heartwarmingly reported that in 1546 King Jan Olbrecht extended the right to distill to all of his subjects, the only problem with this story being that Jan was at the time in no position to exercise that generous impulse, having died almost half a century earlier. Better documented, alas, is the revocation of this general entitlement with the end of the Jagiellonian Dynasty in 1572, when the monopoly on distilling was restored to the aristocrats. Still, a decade later the town of Poznań was already famous as the heart of Polish vodka production, though it soon lost the distinction to Kraków, which in turn ceded the honor to Gdańsk.

From the beginning, distillers all over Poland seem to have added a variety of different herbs to the mash, as a means of disguising the rough congeners that their crude pot stills left in the distillate. But by the seventeenth century they were evidently actively looking for purity, routinely triple-distilling and typically ending up with a spirit of around 70 to 80 percent ABV that they cut with around 50 percent water before selling or consuming it.

In 1693 the Kraków distiller Jakub Hauer published instructions for making vodka from rye, rather than from the traditional wheat. This new grain rapidly took over as the Polish standard until the turn of the nineteenth century, when the potato was introduced alongside it as a base material. The mid-nineteenth-century European potato blight soon caused major disruption; nonetheless, by 1850 some seven thousand small distilleries were operating in the combined Austro-Hungarian and Russian-dominated sectors of Poland, and vodka exports to Russia were becoming a major element of the Polish economy. Calamity struck in the early 1870s, when the Russian authorities imposed a severe excise duty on those exports. Although this episode was an immediate disaster for the Polish vodka-makers, it led ultimately to improvements both in quality and efficiency. Those enhancements, in turn, were greatly helped along by the adoption of Aeneas Coffey's two-column distilling technology, which was a perfect match for the high-octane clear Polish spirit.

One of the greatest rags-to-riches stories of the nineteenth century was that of Pyotr Arsenievich Smirnov. Born into a serf family in the provincial town of Kayurovo in 1831, by midcentury Smirnov found himself in Moscow. There the penniless migrant immersed himself in the liquor business, and by a series of improbable stratagems established himself as a hugely savvy entrepreneur at a time when spirits consumption was booming. After the abolition of serfdom in 1864 he began manufacturing vodka under his own name, using his own adaptation of Theodore Lowitz's charcoal filtration technique. Thanks to the quality of his product, his clever use of newspaper advertising (he also allegedly sent representatives to bars all over Russia to ask for his vodka specifically), his strategic donations to the Russian Orthodox Church, and his company's appointment as purveyor to the Imperial Court of Russia in 1886, by the time Smirnov died in 1898 his vodka dominated the Russian market. His successors fought among themselves but continued, more or less, to flourish in an increasingly restless society until 1914, when Tsar Nicholas II, aghast at the drunkenness of army recruits turning up for World War I, promulgated a dry law that put the Smirnov company out of business (Figure 10.1).

After the 1917 Revolution, Pyotr's son Vladimir was forced to flee Moscow. By 1925 he was in France, where he soon began producing vodka again under the French "Smirnoff" version of his name. Sales were slow to take off, however, in a country deeply attached to its Cognacs and Armagnacs. Eventually, Vladimir sold the American rights to both his filtration technology and the Smirnoff label

Figure 10.1. Pyotr Smirnov.

to a fellow émigré named Rudolph Kunett (né Kunettchenskiy), who duly established a distillery in Connecticut in 1934, the year after Prohibition ended. Unfortunately, business continued to be sluggish in the face of the historical American predilection for whiskeys, and in 1939 Kunett sold out to the Heublein liquor and steak sauce company.

The eve of World War II might not have been the most auspicious moment for the Connecticut distillery to change hands: notwithstanding the perennial popularity of the Bloody Mary (first concocted in Paris in the 1920s), and the invention of the Moscow Mule in Los Angeles in 1941, business remained relatively unexciting. Predictably enough, it continued that way during the early years of the Cold War—although it did get a small bump from contrarian spirits lovers in 1950, in reaction to the patriotic bartenders who had paraded down Fifth Avenue in New York City with banners reading "Down with the Moscow Mule—We Don't Need Smirnoff Vodka." Despite James Bond's best efforts in the early 1960s (as you'll recall, he oddly liked his vodka martini shaken, not stirred), it was not really until the thaw in U.S. relations with Russia in the 1970s, when PepsiCo and the Soviet Stolichnaya vodka company struck a barter deal, that vodka sales really took off in the United States, powered by the immense advertising resources of the manufacturer of Pepsi-Cola.

Heublein, which had at one point been selling its vodka as "Smirnoff White Whiskey—No Smell, No Taste," jumped on the bandwagon, energetically selling its rebranded product as a uniquely clean, pure spirit that left no compromising olfactory traces: "It will leave you breathless." Heublein advertising also implied that drinkers could down limitless vodkas without having to fear a hangover, due to an absence of the congeners that gave other spirits their flavors. A study published in 2010 seems eventually to have borne out this claim, concluding that, although general performance impairment and sleep disruption are unavoidable with the excessive intake of alcohol of any kind, a sample of experienced drinkers reported worse hangovers from bourbon than from vodka.

Whatever the case, by 1967 the dual assault from the two drinks giants had propelled vodka ahead of gin as America's preferred mixer. By 1976 the spirit was outselling all other distilled beverages, including whiskey, and soon Smirnoff was the world's leading spirits brand by volume. There was an eventual ironic twist to this tale of an exiled Russian brand taking over the American market: Under a 1990 deal between Helmut Kohl and Mikhail Gorbachev, Russia was allowed to station troops in eastern Germany at the West's expense. This made hard currency available to Russian soldiers for the first time, and those soldiers proved only too eager to spend their money on U.S.-made Smirnoff vodka. The entire Russian market seemed in prospect; and the U.S. Smirnoff brand (despite a revived Russian Smirnov) would probably dominate it today, had it not been for a variety of complications at the Russian end.

But then, there were always complications at the Russian end. For one thing, Moscow officialdom has traditionally see-sawed between the desire to suppress drunkenness and a dependence on alcohol taxes to pay for government. During Soviet times, the lust for revenue usually prevailed, despite the official anti-drunkenness stance exemplified by the poster in Figure 10.2. But as the Soviet Union began to disintegrate both economically and politically in the mid-1980s, the first instinct of the Kremlin was to try to rescue the system by actively imposing sobriety on the workers who had instinctively turned to vodka as their standards of living deteriorated. In 1985 General Secretary Mikhail Gorbachev accordingly raised alcohol taxes sky-high, and antidrunkenness laws were tightened and enforced in what amounted to quasi-prohibition. There was a temporary uptick in life expectancies, but ultimately the public became even more disaffected,

Figure 10.2. Soviet-era temperance poster.

the economy continued to sag, and cheap and badly made illegal hooch became ubiquitous. This illicit alcohol frequently poisoned the citizenry, who otherwise had only marginally more dangerous perfume or shoe polish to turn to, and tax revenues collapsed.

After power passed in 1991 from the relatively straitlaced Gorbachev to the frankly alcoholic Boris Yeltsin, the state monopoly on vodka production was abolished. Predictably enough, tax revenues plummeted while even more hooch flooded the market. When the Russian economy was privatized in the mid-1990s, vodka became the vehicle by which ruthless opportunists seized the former property of the state from its hapless new owners, as uncomprehending citizens accepted bottles of liquor in exchange for the vouchers the government had given them as their share of the economy. It took well over a decade, and some minor crises, to finally get Russian booze consumption down to somewhere near the European average. This was achieved by dint of measures such as imposing new state controls on production, increasing taxes (again), banning advertising, and strictly restricting availability—although the eventual success of the campaign probably also had a lot to do with the increasingly bourgeois nature of Russian society, which nowadays drinks a lot more wine than of yore.

Meanwhile, both in Europe and in the United States, vodka continued to be a strong player on the spirits market, as brands multiplied and especially as fruit-flavored vodkas began to catch on in the late 1980s. The Russians had always viewed the process of flavoring vodka with some disdain, as a means of disguising flaws in the basic spirit. In Russia a flavored vodka is consequently deemed not worthy of the name, so that a bottle of the lemon-flavored Stolichnaya is labeled "Limonnaya." At one time the Scandinavians also adhered to this strict approach: they have long had their own (mostly) caraway- or dill-flavored vodkas, calling them *aquavit, akvavit,* or *akevitt.* But in the heyday of liquor print advertising (since TV advertising was banned) such purist ideals could not last long. In 1986, Sweden's Absolut vodka (which, through brilliant advertising, was by then the leading import in the American market, and the first avowedly "luxury" vodka brand) launched its pepper-flavored Absolut Peppar in the United States. Emboldened by the immediate success of this novel beverage, Absolut offered Americans its Absolut Citron two years later, and watched its profits soar as the Cosmopolitan became everyone's favorite cocktail. Eventually, Absolut presented America with seventeen different vodka flavors that ranged from grapefruit to raspberry to açai. And although the Swedish company dialed back a little as others piled into a rapidly saturating market, infused vodkas still make up a significant part of American vodka sales.

Still, in vodka as everywhere else, the big profits are to be found at the high end of the market, and soon other vodka producers—including Swedish distillers who trumpeted the virtues of very expensive potato-based vodkas—followed Absolut in coming up with premium and ultra-premium products. Given that the basic element of vodka is its neutrality, the multiplying brands found it necessary to promote these more expensive wares less through their intrinsic gustatory merits than through advertising gimmicks and eye-catching packaging. Do you fancy the notion of vodka made using ancient water recovered from a melting iceberg? Canada's Iceberg Vodka Corporation has your back. Are you entertained by the idea of vodka filtered through a bed of Herkimer "diamonds"? Crystal Head has just the thing—and they will ship it to you in a costly, glittering bottle shaped like a human skull. If you find your vodka more seductive when served from a bottle in the form of a slender young woman, Damskaya from the Russian distiller Deyros has what you need. Not quite Russian enough for you? How about some

Kalashnikov vodka, which comes in a bottle shaped like the eponymous AK-47 semi-automatic rifle and offers a "military strength" ABV of 42 percent? Or maybe you'd prefer a glass of Poland's Jazz vodka poured—you guessed it—from a glass trumpet? And on and on. But in the end, of course, it's all vodka.

Wait just a minute, some will say. Unflavored vodka may be a neutral spirit, and its effects on the mind and taste buds are certainly unaffected by how it is packaged; but those effects *are* influenced by how it is made. Vodka as we know it today was first created in pot stills, and it was produced that way for centuries. But if ever a marriage was made in heaven, it was surely the one between the ideal of vodka and Aeneas Coffey's two-column still, which came into common use after the middle of the nineteenth century (Chapter 2). After all, vodka is all about purity, and so is the column still. A single pass of a 10 percent ABV mash through even the most efficient modern pot still will produce a spirit of only about 35 percent ABV, and the resulting liquid will be full of the chemical congeners that vodka-makers are anxious to eliminate. Multiple distillations will raise the alcohol level in the distillate to a maximum of about 80 percent ABV, removing lots of congeners but still leaving many to be filtered out. Column distillation, in contrast, will hand you ethanol that is pure up to 96.5 percent. As a result, all modern industrial vodkas (that is to say, virtually all the biggest-selling brands) are made on tall continuous stills, using a technology borrowed ultimately from oil refining.

Indeed, the basic spirit in your vodka bottle is rarely made by the manufacturer whose logo appears on the bottle label. Most of today's big American vodka producers are more like compounders than distillers: they buy their neutral grain spirits in vast quantities from bulk producers such as the Chicago-based Archer Daniels Midland Company, or the Grain Processing Corporation of Muscatine, Iowa, at a purity of around 95 percent. Then they filter, dilute (usually with processed water), bottle, label, and distribute them, perhaps adding trace flavors along the way (or more overtly flavoring their infused vodkas). It is largely during this last step that any differences between brands may enter the picture. A 2010 study attributed differences among various brands to slight variations in hydrogen bonding between the water and ethanol molecules; and, logically, gustatory differences might be related to those trace additions. It is up to the consumer to judge whether such differences are meaningful, or even if they are detectable by the individual palate. In any event, given the common origins of most of the neutral grain spirits on the market, it seems likely that the cheapest industrial vodkas are the most pristine.

But then, of course, there are the craft distillers. These artisans make their vodkas in pot stills, either from scratch (often blending batches from different grains) or from purchased column spirits. Cynics may wonder what is the point of going to all this trouble, if the goal is a neutral product that is more cheaply and easily made using giant column stills. But the craft distillers fervently retort that all aspects of the process are crucial in determining the outcome. These include the water (both for distillation and dilution, and usually from wells or springs, though occasionally from conditioning machines); the base from which the spirit is distilled (unlimited in the case of vodka: a couple of craft distillers use only whey from cow's milk, which is said to impart a creamy mouthfeel); batch size; the material from which the still is made; how the distillate is filtered; and maybe above all, the imponderable element of the distiller's expertise.

Only the consumer can judge whether the differences, if any, are worth the added expense. Chances are that many cocktail drinkers will answer this question in the negative, while those who prefer their vodka straight, or on the rocks with a hint of lemon peel, will tend to agree. And then, of course, "craft" is also a matter of scale. There is no formal definition, though most of those who recognize themselves as craft distillers hold "licensed distiller" certification, which entails producing a maximum of 750,000 cases a year and distilling on premises. Even so, the largest American manufacturer of "handmade" vodka (which admittedly began two decades ago as a one-man outfit with a single small still) starts with industrial spirit and bottles almost eight million cases of finished vodka every year (second only to Smirnoff).

So what's next for a spirit that is famed for its neutrality? Although vodka is a fairly new arrival on the international scene, it is nevertheless the ultimate international drink because it bears no traces of local terroir (even if the quality of ingredients is still important, especially in craft products). There is no way even the most knowledgeable connoisseur can tell where a particular glass of vodka originated. In most places, vodka does not have quite the same deep cultural resonances and overtones of social bonding that it has in Russia and Poland or even in Scandinavia; yet this elusive spirit is nonetheless a deeply entrenched component of the international drinking scene, and it seems destined to continue that way. Despite the moans of those who believe that a cocktail should reflect the character of the base spirit, neutral vodka will remain the ultimate mixer. In better economic times, consumers will probably gravitate toward premium brands, whereas in recessions there is plenty of flexibility for them to move down-market without sacrificing the basic nature of their tipple. But it is hard to envisage any point in the economic cycle at which drinkers will entirely lose their affection for what is, after all, not only the most adaptable spirit in the library, but also the cheapest to produce. What's more, its vaunted neutrality largely protects vodka from the vagaries of fashion. Chestnut barrels or long sea voyages will never be a factor in selling this spirit.

Still, given the omnipresent human craving for novelty, distillers and their advertisers will probably continue to promote any innovation as a way of reeling the punters in. Among such novelties, new or more exclusive raw materials (someone's already made a cactus vodka), innovative filtration materials (at this point, celebrities' underwear wouldn't be surprising), and new flavors (as if cookie dough weren't enough) are the most obvious. But who knows what else might be in store? The possibilities are limited only by the bizarre human imagination. Nonetheless, it's reasonable to expect that, whether you enjoy it in a fantastical ice palace carved into a Finnish glacier, in a freezing "vod-box" inside a trendy Beverly Hills restaurant, or in your own cozy living room, the vodka itself will stay essentially the same.

11

Tequila (and Mezcal)

Ignacio Torres-García, América Minerva Delgado Lemus,
Angélica Cibrián-Jaramillo, and Joshua D. Englehardt

Siembra Valles Ancestral Tequila Blanco is truly unique. Produced artisanally by the Rosales family of Destilería Cascahuin in El Arenal, Jalisco, in collaboration with the Vieyra family of Mezcal Don Mateo de la Sierra of Pino Bonito, Michoacán, this spirit offers a return to tequila's origins. Its creation starts with sustainably harvested, bat-pollinated agave, processed using traditional techniques: the piñas are roasted in an earthen pit oven for almost five days, hand-crushed with wooden mallets, and fermented in an oak barrel. It is distilled first in an old-fashioned copper alembic, then transferred to an even older, Filipino-style pinewood still for the second distillation, before resting for six months in glass demijohns plugged with maize cobs. These ancestral methods result in a spirit that is close to the traditional mezcals from which all tequilas derive. Traditional pit-roasting, in particular, gives this tequila a noticeably smoky aroma, complemented by a nose of honey, citrus, and celery. It has a smooth, silky mouthfeel with the sweet, earthy flavors of roasted agave, accented by complex notes of black pepper, oak and pine, minerals, and spices, as well as a light and gentle finish—somewhat surprising for a 100-proof spirit. As with any 100 percent agave tequila, it is best sipped neat from a caballito or shot glass. Don't even think of turning this gorgeous, timeless spirit into a margarita.

Very few beverages, alcoholic or otherwise, are as strongly associated with their country of origin as tequila is. But although legend, myth, and popular history venerate tequila as the emblematic Mexican spirit, its origins and history are controversial. In fact, alcoholic drinks based on the many species of agave (especially *A. tequilana* Weber var. *azul*), and other native succulents such as maguey (*A. salmiana; A. vivipara*) and sotol (*Dasylirion wheeleri*), have a history far deeper than that of the modern nation of Mexico.

Indeed, agaves and their botanical cousins appear to have been fundamental in the development of indigenous Mesoamerican and Aridoamerican cultures, both as a basic foodstuff and as the basis for specialized drinks used in rituals. Archaeological evidence indicates that plants in the *Agave* genus have been important for at least the past ten thousand years, and evidence for the cooking of agave in pit ovens—the first step in the tequila-making process—has been found at numerous archaeological sites. The sweet food produced by roasting agave in earthen ovens was known to the Nahuatl-speaking populations of central Mexico as *mexcalli* (hence *mezcal*). In conjunction with other wild and managed resources, this and similar foodstuffs likely constituted a considerable part of the pre-ceramic and pre-agricultural diet of many regional groups. The pantheon of the later Mexica culture—more commonly known as the Aztecs—included the goddess Mayahuel, a female deity associated with maguey and more general notions of abundance, fertility, and nourishment.

Coincidentally, Mayahuel was also associated with the discovery of pulque, the pre-Hispanic alcoholic beverage (and distant cousin of tequila) made from fermented maguey sap—said to be the goddess's blood—and consumed as part of religious festivals. Perhaps a more evocative origin story for pulque involves the *tlacuache* (opossum), which clawed his way into the maguey heart to extract its fermented juice, thus becoming the first drunkard. In any case, it is clear that agave was one of the most sacred and important plants in the ancient Mesoamerican world, having a privileged role in the mythology, ritual, foodways, and economies of numerous indigenous societies.

Some scholars have argued that distillation existed in the Mesoamerican world prior to European contact, although this is highly speculative. It is more generally agreed that agave-based distillates result from a combination of traditional practices for cooking agave stems and Asian distilling methods introduced by Philippine and Spanish migrants during the colonial period (1521–1810). Local knowledge of agave's properties and preparation was undoubtedly important for the production of tequilas and mezcals, since traditional techniques result in the accumulation of highly rich, fermentable sugars in the agaves. Although there is some controversy over the exact timing, it is most likely that distilling was introduced to Mexico between 1570 and 1600 by Philippine immigrants to Colima who arrived on the "Manila Galleons." In addition to stills, these immigrants brought with them the coconut palm, and in fact the first liquors distilled in Mexico were not agave-based, but were rather from coconut wine (lambanóg) in the Philippine tradition. (For more on lambanóg, see Chapter 2.)

Also in the late sixteenth century, Spanish migrants arriving from Europe introduced the Arab alembic still and the external condensation distillation technique. Both Spanish and Filipino tools and techniques spread throughout the Spanish colonies that now comprise Mexican territory. Subsequently—according to legend, when the Spaniards' supplies of brandy were exhausted—these technologies were adapted for use in distilling the fermented juices and fibers of cooked agave, thereby producing one of the New World's first indigenous—or more properly, *mestizo*—spirits. Certain techniques became more popular in particular regions: for example, traditional stills found in the states of Oaxaca, Michoacán, and Jalisco have traits common to Philippine ones, likely owing to the expansion of this distillation culture from its origins in the Colima lowlands to neighboring areas of highland west Mexico and along the Pacific coast. Nonetheless, by 1621 the *Description of Nueva Galicia* (modern Jalisco state) suggests the presence of both Philippine and Arab stills in this region, which is known as the heartland of modern tequila production.

Tequila itself was likely first produced in the sixteenth century, in the Altos and Valle de Amatitlán regions of Jalisco, home to Tequila, the town that ultimately lent its name to the drink. The rich, volcanic soils in these areas were—and remain—particularly well-suited for cultivation of *Agave tequilana.* Don Pedro Sánchez de Tagle, second marquis of Altamira, is known as the "Father of Tequila" for establishing the first tequila factory at his Hacienda Cuisillos around 1600 (its exact location is still a subject of dispute), and the first "official" mention of the spirit, in 1616, refers to the "mescal wine of Tequila," suggesting that its production already represented an important source of income for the colonial authorities. Significantly, almost all early records of agave-based spirits refer not to "tequila," but to "mescal wine."

At this point, we must address a contentious subject: the distinction—or lack thereof—between tequila and mezcal. Some insist that all tequilas are mezcals, whereas others argue that tequila is in its own category of spirit. To some extent, saying that tequila is a type of mezcal is both true and false: True, in the sense that tequila was indeed originally baptized as a mezcal, the "mescal wine of Tequila." But false, in that mezcal literally means "burnt maguey" (from the Nahuatl *mexcalli: metl,* or "maguey," and *izcalli,* "burnt or cooked with fire"), which is no longer true for all tequilas. Thus, although tequila was indeed "originally" recognized as—and technically remains—a type of mezcal, modern production processes are so different from the traditional methods that today's tequilas are nothing like their original versions.

Agave-based spirits, all of which are generally referred to as *mezcales,* are currently produced in twenty-four Mexican states, from around fifty-four agave species. Each spirit is a unique representation of the cultural identity, available resources, and distinct technological expressions of its producers. From the specific agave varieties chosen, the properties of water and types of wood used in roasting and distilling, and the various modern and traditional skills and technologies employed individually and collectively among mezcal-making communities across discrete geographical, political, and ethnic divisions, it is clear that these spirits are part of an extraordinarily diverse and complex food web with a centuries-long history.

The primary ingredients (fermented agave juice and water), alcohol contents, and basic flavor profiles of the various tequila and mezcal varieties are similar. Apart from the ethanol, both are composed of water and traces of higher alcohols, methanol, aldehydes, esters, and furfural, and, in mezcal, of fibers used in the distillation process. The alcohol content in traditional mezcal ranges from 45 to 52 percent ABV, while in tequila today it is usually less than 45 percent. Most important to the unique tastes and properties of distinct distillations is the local fermenter microbiome, which includes yeasts and bacteria common to other fermentation processes, such as those for wine.

The primary difference between mezcals and tequilas is described in the *Norma Oficial Mexicana* (NOM), a set of compulsory legal regulations applied to all facets of tequila production. It stipulates that in order for a spirit to be labeled "tequila," it has to be made from the blue *Agave tequilana*. Mezcal, on the other hand, may be fabricated from as many as forty types of other agave species. Production processes also differ, from the cultivation of the plant itself to the cooking, fermenting, and distillation, with tequila production being more industrialized, and mezcal production tending to be more traditional. For example, in tequila production agave *piñas* (stems) are usually cooked in above-ground steam ovens or autoclaves, whereas in preparing mezcal, wood roasting in conical, rock-lined pit ovens is preferred. These differences impart to mezcal a more intense, smoky flavor of roasted agave, whereas tequila's taste is less smoky and slightly sweeter. Another distinction lies in the separate appellations of origin regarding tequila (AO) and mezcal (AOM): the AO specifies that the spirit can be produced only in certain municipalities within five Mexican states (Jalisco, Michoacán, Nayarit, Guanajuato, and Tamaulipas), whereas the AOM is broader, permitting production in twelve states.

The distinction between tequila and mezcal may appear mainly semantic, but the differences in the production processes are critical. Further, the technical and legal distinctions are relatively recent constructs—the AO was created in 1974, while the first iteration of the NOM dates to 1978 (building on an earlier 1964 law). In fact, use of the term "tequila" to refer to something somehow distinct from mezcal started only in the late nineteenth century, following expanded exportation to the United States. It is also important to note that the AO and NOM are overseen by the powerful Tequila Regulatory Council (TRC) and respond primarily to market pressures and the economic and political interests of large-scale producers of "official" tequila. For example, the original AO allowed production only in Jalisco, where most tequila companies are based. In this sense, the history of tequila is both fundamentally entwined with that of mezcal, its direct forebear, and fraught with the political and economic factors involving the recent exploitation of indigenous and mestizo knowledge by multinational corporations—even if some of these corporations (for example, Cuervo) remain family-owned. This profit-seeking often occurs to the exclusion and detriment of communities that have traditionally produced agave-based spirits for over four hundred years.

Although many regard tequila as emblematic of the modern Mexican nation-state, we authors, all of us Mexicans living in tequila- and mezcal-producing states, would consider mezcal more appropriate in this role, having a closer connection to indigenous culinary heritage, and being a truer distillation (if you will) of Mexico's complicated history. We might also note that all of us generally prefer mezcal to tequila. Indeed, mezcal is a unique and complex spirit with infinite variations, worthy of its own chapter. Nonetheless, in the remainder of this chapter we will focus primarily on tequila, Mexico's ambassador spirit.

The production of tequila involves seven basic steps: harvesting, cooking, shredding/mashing, fermentation, distilling, aging, and bottling. Agaves flower only once, at the end of their lifetimes, so the *jimadores* who cultivate these plants take care to ensure that they will not flower early, lest the accumulated sugars turn into flowers, nectar, and seeds. They recognize the ideal moment to harvest the ripe, sugar-rich agave stem head, or piña—usually just before inflorescence, approximately five to eight years after planting. Piñas, which can weigh up to a hundred kilograms, are extracted using a specialized tool known as a *coa,* after which they are transported to ovens for the second production step. It is said that highland agaves yield sweeter, more floral tequilas, whereas those grown in the lowlands impart an earthier flavor.

Roasting the agave piñas breaks down complex carbohydrates into smaller sugars that can ferment easily. Traditionally, cooking was done in a pit oven, and piñas were roasted over high-density firewood from various trees that are characteristic of the region of production (often species of *Quercus, Lysiloma,* and *Prosopis*). The traditional process allowed aromas unique to each tree species to penetrate the piñas—akin to the smoky flavor imparted by grilling meats over mesquite charcoal—thereby developing a distinct organoleptic profile for each batch. In the more industrialized, large-scale production processes that have developed since the late nineteenth century, however, piñas are steam-cooked in large masonry furnaces (*hornos*), usually employing fuel oil to generate vapor. Piñas are usually cooked for fifty to seventy-two hours at temperatures ranging from 140 to 200 degrees F (60 to 95 degrees C); this slow roasting softens the agave fibers and prevents the piña from caramelizing. Some large distilleries use metal autoclaves that cook piñas in twelve to eighteen hours, and this industrial-level preparation significantly decreases the smoky aromatic profile that is such an important component in traditional agave-based distillates.

After baking, the roasted piñas are either shredded or mashed, washed with water, and strained to extract the syrupy juice. Traditionally the mashing or shredding was often done using a large stone wheel mill known as a *tahona,* usually powered by pack animals; it could also be accomplished by smashing the piñas with wooden mallets. Modern distilleries generally employ mechanized shredders and crushers. The extracted agave juice is then separated from the pulp fiber (*bagazo*) of the shredded piña and placed in large wooden or stainless-steel vats for fermentation. The fermentation process takes between one and seven days, depending on factors such as ambient temperature and the size and composition of the vats (width, material, texture, and so on) that affect aspects such as temperature consistency and flavor.

In traditional production, fermentation is spontaneous (as in lambic beer) and depends on the diversity and abundance of wild microorganisms that naturally occur locally, with yeasts—usually *Saccharomyces cerevisiae*—as the main drivers of the process. Tequila Herradura claims to still use a natural fermentation process. But during fermentation other commercial producers add yeasts that are bioengineered or selected to homogenize flavor and ensure a consistent organoleptic profile among batches, as well as to eliminate unwanted microbes. Some producers introduce bagazo into the fermentation process to impart a stronger agave flavor to the final product. Water is also key to fermentation, since agave sugars must be diluted by about 15 percent in order for fermentation to occur. The water must be free of chlorine, to avoid killing the microorganisms that promote fermentation, and its quality and source are important. Spring and well waters have distinct mineral profiles that in turn define the aromatic and flavor profiles of the distillate; for this reason many producers have their own water source.

The low-alcohol wort, or *mosto* (also called *tepache* or *tuba*), that results from fermentation is then distilled. The NOM requires at least two distillations: the first produces unclassified *ordinario,* while the second results in *blanco* (white) or *plata* (silver) tequila. The first distillation takes around a hundred minutes at approximately 200 degrees F (95 degrees C) and converts the mosto from 4–5 percent to 25 percent ABV. The second distillation lasts between three and four hours, and results in a spirit of around 55 percent ABV, which is then diluted with demineralized water to the desired proof (76–80 proof or 38–40 percent ABV). Very rarely, a third distillation may be undertaken, although many say this removes the agave flavors from the spirit. All tequilas have a similar alcohol content, although generally those sold in Mexico are closer to 38 percent, while export varieties are at 40 to 50 percent. Tequila may legally be up to 100 proof, although that is uncommon, and the minimum ABV is 35 percent. If the tequila is to be aged, some distillers will leave the distillate at a higher proof to compensate for evaporation during the aging process. Traditional distillation methods are remarkably diverse, and they vary according to local technologies and customs. Industrialized tequila production, however, generally employs alembic pot stills of the Arab or Philippine varieties, although over the past several decades larger operations have adopted column stills for expanded production. Smaller producers still use batch stills descended from those originally introduced in the sixteenth century. Once the wort has been distilled twice, it may legally be called tequila—and consumed as such at that point. The sixth step of the production process—aging—is optional; and the final one, bottling, varies according to the manufacturer.

The NOM establishes five legal varieties of tequila: *blanco/plata, joven/oro, reposado* (rested), *añejo* (aged), and *extra añejo* (vintage). Tequila blanco is the product of the second distillation and is the most common and "original" form of the spirit. Some aficionados insist that blanco is the purest form of tequila, since it captures a more robust agave taste than more "refined" varieties. Joven is simply tequila blanco with added color and flavorings to make it look aged and to smooth out the harsh taste that some may find in it. The resposado, añejo, and extra añejo varieties are all aged in wood containers (usually made of American, French, or Canadian oak). The interaction between tequila and the aging barrel leads to chemical reactions among the tequila components and the creation of new compounds, depending on the properties of the wood itself. This process can impart further subtle notes to the spirit, creating richer, more complex flavor profiles and aromas and mellowing the harshness of the alcohol.

Tequila reposado is aged for between two and twelve months in either small barrels or large casks of up to twenty thousand liters. This variety accounts for over 60 percent of tequila sales in Mexico. Añejo and the recently created extra añejo varieties are aged in government-sealed barrels with a maximum capacity of six hundred liters (usually closer to two hundred) for a minimum of one and three years, respectively—and in some cases as long as ten years. As tequila ages it acquires darker colors, and the wood of the barrels imparts more and more unique flavors. Although aging could theoretically continue for decades, many insist that tequila is at its best after four or five years.

Two basic types of tequilas are recognized by the NOM: 100 percent agave, and *mixto* (mixed). The 100 percent agave label indicates that no other sugar was used during fermentation to create the alcohol, and that it is made purely of *Agave tequilana* Weber var. *azul.* Consequently, any variety of 100 percent agave tequila will possess more body, flavor, and a stronger agave aroma. By law, all 100 percent agave tequilas must have a NOM identifier on the bottle that includes a government-assigned distillery number. If the label on the bottle does not specify 100 percent agave, the tequila is by definition mixto. Mixtos are made from a minimum of 51 percent agave sugars, with other sugars (such as glucose and corn sugars) making up the remaining 49 percent. They can also legally contain ingredients such as artificial colorings, glycerin, sugar-based syrup, and oak extract, and may be bottled outside of Mexico. Mixtos are never labeled as such; rather, they are simply called "tequila." Mixtos were first produced in the 1930s and are less expensive to make than pure agave tequila. Although there are far more mixto brands than 100 percent agave brands currently available, especially outside Mexico, most connoisseurs agree that "100 percent agave" on the label indicates better quality, flavor, and purity, as well as fewer of the additives and congeners that are responsible for the notoriously horrible hangovers that can result from drinking too much low-quality tequila.

According to the TRC there are more than 1,400 registered brands of tequila currently on the market, distilled by over 150 producers. There are fewer distilleries than tequila companies, in part because some companies lease out their premises and equipment to smaller distillers, allowing them to produce without significant capital outlays. Some larger estates, too, produce limited batches that employ traditional production methods and that arguably approximate more closely the "soul" of the original mezcal wines of Tequila. There is thus a tremendous variety of brands of tequila available, crafted to conquer diverse tastes—not to mention hundreds of commercial varieties of mezcal. There is an agave-based spirit of Mexican origin for every palate.

When distilling reached Mexico, local agave-based spirits entered into competition with grape wines and imported European liquors. Responses by the colonial authorities alternated between taxation and prohibition. Although taxes had been imposed on mezcal production since 1608, in 1742 the government of Nueva Galicia attempted to prohibit and impose fines on the making, marketing, or "excessive" consumption of "mescal wine." And in 1785, King Charles III instituted a total ban on the production of all agave-based alcohols to protect imported Spanish liquors. This ban did little to hamper mezcal making, and often spurred production as a symbol of resistance and local identity. A decade later, following the death of Charles III, the Spanish Crown realized that the popularity of mezcal could make taxation more profitable, and reversed course. In 1795, King Charles IV provided the first official license to produce "mescal wine in the Tequila region" to José María Guadalupe de Cuervo. Tequila began to spread throughout New Spain, as well as the Crown's northern colonies, due largely to the boom in northern mining industries. Taxes on mezcal production would eventually be used to construct the University of Guadalajara in the late eighteenth century.

Until Mexican independence in 1821, mezcal production remained relatively small-scale, limited to a handful of family-owned *tabernas* and distilleries. Following independence, imports of Spanish wines and liquors decreased, and sales of mezcal bloomed. During the rest of the nineteenth century new technologies emerged, and localized production expanded into large-scale *fábricas* to satisfy a growing market. This expansion was periodically interrupted by internal and external conflicts such as the Mexican-American War (1846–1848), and the War of the Reform and subsequent French intervention (1857–1867), although these episodes also introduced agave spirits to new audiences. Around 1850, the preferred method for baking agave piñas became above-ground masonry ovens rather than earthen pit ovens. Although producers adopted this change only gradually over the following seventy years, this shift signals the point where tequila and mezcal began to diverge.

In 1870, several large producers in Jalisco asked for—and received—permission from the Mexican government to officially name their versions of mezcal *tequila,* after the town of the same name. The first exports of tequila were made to the United States in 1873—although there is some debate as to whether the Cuervo estate or Don Cenobio Sauza (often credited with the initial suggestion that *Agave tequilana* Weber var. *azul* was the best agave for tequila) first exported the spirit. In the same year, tax records note the name tequila for the regional mezcal, as well as the export of mezcal and other Mexican spirits to England, Spain, France, and New Grenada, thus beginning tequila's long romance with the rest of the world. After a "mezcal de Tequila" won an award at the 1893 Chicago World's Fair, tequila-makers and the Mexican government definitively dropped "mezcal" from the name.

By the end of the nineteenth century tequila exports were growing steadily, and many of the "major" distilleries and estates had been established in Jalisco. These included Los Camichines (1857), Destiladora de Occidente (1860s), and Tequila San Matias (1886). Further infrastructural and technological improvements (rapid expansion of rail networks, power mills, presses, and grinding equipment) all contributed to increased demand, production, and the spread of tequila into new markets. This growth persisted after the turn of the century, when the sale of bottled (as opposed to barrel) tequila was introduced by the Cuervo estate, and this upswing generally continued through the first half of the twentieth century—despite social and political upheavals such as the Mexican Revolution (1910–1920) and the Cristero Rebellion in Jalisco (1926–1929). Prohibition in the United States also boosted tequila's popularity north of the border.

In the 1930s and following decades, agave shortages driven by increased demand forced the Mexican government to relax production regulations, thereby giving birth to mixto tequilas, which became quite popular in the U.S. market. At this time the Mexican government also established a series of laws related to the production and taxation of tequila, culminating in the Industrial Property Law of 1942 that laid the groundwork for tequila's appellation of origin. At some point between 1935 and 1942, tequila's star cocktail, the margarita, was invented, although the stories of its year and place origin differ—some say it originated in 1935 or 1938 at a bar just outside of Tijuana, while others claim it was invented in 1942 in either Ciudad Juárez or El Paso, Texas. During World War II demand increased yet again, due to the scarcity of European spirits in both Mexico and the United States. Tequila production grew 110 percent between 1940 and 1950, and doubled yet again by 1955, serving as an incentive for foreign investment and further commercialization. This trend continued through the 1960s as tequila's popularity continued to increase, particularly as a neutral spirit for mixology.

Further expansion followed the 1974 "Declaration for the Protection of the Appellation of Origin Tequila." At this time, the Cuervo and Sauza estates were producing 60 percent of all tequila, although in the "tequila boom" years of the 1980s and 1990s the industry received increased governmental support and the industry became truly globalized through foreign investment and partnerships between some of the larger tequila producers and multinational corporations. The growing popularity of Mexico as a tourist destination resulted in an expansion of brands and different styles that were marketed not only in Mexico, but all over the world. The TRC was established in 1994, and in the same decade U.S., Canadian, and European laws recognized tequila as a uniquely Mexican spirit. (Coincidentally, mezcal was granted its own AO in 1995.) More recently, distilleries in South Africa, Japan, and Spain have attempted to produce "tequila" outside of Mexico, although these drinks are not labeled as tequilas.

Although the United States has been the largest consumer of tequila for many years, by 2000 internal sales and consumption within Mexico almost equaled exports (with 84 million liters exported and 72 million liters drunk in Mexico). From the turn of the century through 2019 tequila production almost doubled, from 182 to 352 million liters, and exports increased by a factor of 2.5—from 99 to 247 million liters. These increases have been driven almost exclusively by higher demand for 100 percent agave tequila. According to TRC data, production of this variety has increased by a factor of ten over the past twenty years and exports have expanded by a factor of thirteen. With a trade deal concluded in 2019 that allows for exportation to China, these trends show no signs of slowing. As one commentator put it, tequila is taking over the world, one glass at a time.

The recent worldwide appeal of tequila has profoundly affected the culture surrounding its drinking. While it once held a reputation as a "party" drink, shifting tastes have changed how the spirit is prepared and consumed. In some ways it has reverted to more "traditionally Mexican" tastes and conceptions regarding its enjoyment. Within the country of its origin, tequila is embedded in many aspects of Mexican culture, from the ancient heritage of indigenous agave production to contemporary national pride, and its consumption befits its iconic stature. Mexicans generally prefer their tequila neat, served in a shot glass (*caballito*), and sipped rather than slammed. The *bandera* presentation, traditionally a shot of tequila blanco with accompanying glasses of lime juice and spicy, citrusy *sangrita*, is also popular. The lime juice and sangrita act not as "chasers," but rather as palate cleansers, and are sipped in alternation with the tequila. We regard shooting tequila much as a Scot might look on downing a glass of eighteen-year-old single malt in one gulp.

Much of tequila's "rowdy" reputation, in fact, is due to the "lick it, slam it, suck it" drinking style generally popular among the younger crowd. Like most cocktails associated with tequila—including the famous margarita—shooting with salt and lime supposedly masks or mitigates what some consider tequila's "harsh" taste, or at least gets the pain out of the way quickly. But any harshness is most often due to low quality. Recall that until recently most tequila exports were of the lower-grade mixto variety, and it is still estimated that almost 70 percent of tequila sold in the United States is mixto for use in margaritas (that country's most popular cocktail). In the end, there is no escaping the pain that bad tequilas will inevitably bring; as one reviewer notes, bad tequila is so bad that even ice won't mask it.

Despite the existence of hundreds of cocktails associated with tequila, over the past few decades sipping tequila has found increasing favor worldwide as drinkers have discovered 100 percent agave and "premium" tequilas. This trend is most evident in the exponential increase in the production and export of 100 percent agave brands and varieties over the past twenty years. And while for some abominable reason some may still insist on shooting a caballito of Gran Patrón Burdeos añejo (priced at around five hundred dollars a bottle)—an act that, quite frankly, offends Mexican sensibilities on many levels—tequila has recently moved into the pantheon of world spirits that are meant to be savored and enjoyed neat to experience the rich, subtle complexities of nose, flavor, and finish. Likewise, novel cocktails have sought not to disguise tequila's flavor, but to play off its nuances. Such developments have contributed to a broader appeal among cocktail lovers and aficionados, and to fundamental shifts in tequila consumption practices.

The outlook for tequila, like its history and its country of origin, is complicated. Distillers' clever blending of modern technology and ancient knowledge has given the spirit an authentic, almost mythical aura that has helped to drive its recent surge in popularity. Tequila retains an intimate association with its source—the agave—and the indigenous traditions associated with the cultural uses of this plant. Moreover, the production of tequila and other mezcals still hinges on the experience and knowledge that traditional producers have transmitted through generations, as an intangible cultural patrimony. Sadly, however, this heritage is in danger of being lost, as is the biocultural diversity that once characterized agave-producing regions and landscapes, due to tequila's dramatic, worldwide increase in popularity over the past century. It's exactly because the world now knows and loves tequila that its future is at risk.

Tequila has evolved dramatically over time in its ingredients, its production processes, and even its alcohol content. The changes have resulted in a fundamental reformulation in the chemical composition of the ancestral mezcal from which it derives, to the point where tequila is hardly recognizable as a descendent of traditional mezcals. Almost all these changes have been driven by market forces: greater demand leads to increased production, regardless of the consequences that such transformations may bring. The endorsement of such modifications by the NOM, TRC, and AO, among others, may provide legal justification; but at its heart these groups are simply leveraging the cultural and historical integrity of this iconic beverage for commercial purposes. Consider, for example, the supposedly "Mexican credentials" of tequilas produced and marketed by large, multinational corporations such as Beam Suntory (the U.S. subsidiary of a Japanese holding company, and current owner of Sauza Tequila).

Worse, most artisanal production methods are disappearing because of the great physical effort and time required. Some boutique distilleries have emerged, offering premium 100 percent agave tequilas produced via traditional techniques; Los Abuelos, for example, uses a vintage *tahona*—a large, volcanic stone wheel—to crush cooked agave piñas. Their production capacity is, however, insignificant in comparison with large commercial distilleries. Moreover, for economic and tax reasons these premium products may be available only outside Mexico, that is, removed from the land of their origin and separated from the inheritors of the cultural traditions that created them. Some have questioned how much longer tequila can remain a truly Mexican industry and a Mexican icon if the country loses control of a large portion of its producers. Bearers of traditional knowledge may no longer reap the benefits that their centuries-long heritage should provide.

In terms of genetic and biological diversity, tequila's explosion in popularity has fundamentally altered the landscapes in which agaves are cultivated. As we noted earlier, agave-based spirits are distilled from approximately fifty-four different species, of which only four are currently intensely cultivated and linked to large industries (the vast majority are either extracted directly from wild populations or managed in traditional agroforestry systems). Over time, constant production and artificial selection of specific characteristics like sweetness and size have resulted in distinct domesticated varieties that did not exist in nature. Whereas at least ten different varieties of agave were once used to make "mescal wine of Tequila"—each with distinct characteristics that imparted unique flavors and aromas—now the monocultural production of *A. tequilana* Weber var. *azul* dominates, and most of the historical agave species in tequila-producing regions have practically disappeared.

Tequila's expanded appeal has thus started a vicious cycle that prioritizes the increasingly intensified cultivation of only one species for commercial use, thereby contributing to the degradation of the ecological systems in which agaves grow. Further, current industrial production of blue agave uses emerging biotechnologies capable of cultivating thousands of hectares by using genetic clones of two or three plants, and agrochemicals. Although these innovations indeed expand the production capacities of the tequila industry, they do so to the detriment of traditional—and more sustainable—agroecological management practices. Why should we care as consumers? Because the homogenization of landscapes and ecosystems, as well as the genetic configuration of these plants, can lead only to the homogenization of the drink itself, paradoxically threatening the unique flavor profiles and distinct qualities of the 100 percent agave tequilas that have driven the most recent boom. Expanded industrialization has also led to higher cultivation costs, which in turn increases the price of *A. tequilana,* often pricing out all but the largest producers. Although most agave species cultivated for mezcal production have thus far remained commercially anonymous, the expanding popularity of mezcal—still produced in traditional ways—is also cause for concern. We worry that a similar trajectory might play out here.

In a cruel irony, large-scale efforts at preservation have actually reinforced the vicious cycle. For example, the tequila AO and NOM have historically been modified according to the desires of tequila-makers, who are themselves responding primarily to market pressures. Alterations to these legal instruments have allowed for the intensification of agave production, its expansion into new territories unrelated to its historical cultivation, and the incorporation of new technologies into traditional processes.

Over its long history, tequila has been drastically transformed from a product of traditional knowledge and practices into a globalized commodity that has profoundly changed the biocultural landscapes where it is produced. Millions of hectares of dry seasonal forests—wild, highly biodiverse zones in which agave, corn, beans, and squash were once intercropped in the remnants of wild ecosystems—have been turned into intensively cultivated plantations that are highly dependent on agrochemicals and bioengineering. The result: pollution, soil degradation, and loss of genetic diversity. Although in 2006 UNESCO declared the agave landscapes of Jalisco a "Cultural Heritage of Humanity," to our minds they are anything but. Rather, they are "blue deserts"—degraded territories where only blue agave and agroindustry remain of what once was a thriving symbiosis of nature and culture.

Fortunately, there are efforts under way to counter this biocultural degradation, including some aimed at protecting the ecosystems in which the agaves grow. Many of these initiatives have just begun, so only time will tell if they will be effective. But tequila producers have proven to be very responsive to market forces, which means that consumers can be powerful allies in promoting both the traditional integrity of the spirits and the conservation of the cultural and natural heritage that underpins their production. In order to continue enjoying quality tequilas and mezcals, and to ensure sustainability for the cultures from whose millenary traditions and homelands this spirit is derived, we tequila drinkers must stay informed and demand quality from the field to the bottle. Because in the end, it is we who have the last say—or rather, the last sip. ¡Salud!

12

Whisk(e)y

David Yeates and Tim Duckett

It was a bucolic autumn afternoon, and we were relaxing on leather chairs in Tim's Tasmanian distillery. While we were musing about the rich, complex flavors in some of his cask-strength beauties, Tim remarked that a great whisky should taste like the shape of a sauropod dinosaur. He clarified: A good whisky tastes fine and smooth on entry, like the small head and narrow neck of a sauropod. It should then expand in layered complexity on the palate, like the great body of a *Brontosaurus,* and finish velvety and long, like the great sauropod's tail. If the taste-shape is like a *Tyrannosaurus,* something is wrong:

too many harsh notes on entry, and a short, brutish finish. Likewise, if the taste evokes the shape of a *Stegosaurus*, with huge, angled plates all the way down its back and stabbing horns on its tail, the spirit is too rough and unpalatable. This being Tim's place, we tasted nothing but sauropods all afternoon.

In its most basic form, whisky is a spirit distilled from grain. In making it we essentially extract the active components from a primitive kind of beer, rather like those made before hops began to be added about a thousand years ago. Whisky also differs from other grain-based spirits in that it is aged, often for long periods, in oak barrels. Contact with the chemicals in wood smooths the rough palate of the raw spirit, adding flavors and color. Before refrigeration was available, the distillation of grain was a way of indefinitely preserving the calorific energy of last summer's produce in liquid form. While the production of whisky nowadays is a complex process, it is easy to imagine how it evolved by trial and error, with steps like barrel maturation added through necessity, coincidence, and happenstance. One of these complexities is the name: for simplicity's sake we use *whisky* throughout the chapter except for Irish and American whiskey, but a full discussion of the "whisky" versus "whiskey" spellings appears in Chapter 3.

Whisky offers our senses one of the most complex flavor experiences of any spirit, influenced primarily by the grain used (for example, barley, rye, corn, wheat), the terroir (including geology, soil, vegetation, and microclimate), the water used in various parts of the process, the fuel used in drying the malt, the type of yeast used in fermentation, the shape of the still, the type of maturing cask, the weather during maturation, the length of maturation, and the kind of finishing cask. In general, the closer the influence to the end of the process, the greater its effect on the flavor.

The precursor of Scotch whisky was probably the heather-flavored ales made from barley malt that were brewed by the Picts and their ancestors. Evidence of this brewing dates back to before 2000 BCE. In the later Middle Ages, stills likely arrived from Ireland, where monks had in turn obtained them from the European continent. Initially those stills were used in monasteries for medicinal purposes only, but soon it became common for households to distill beer into a stable alcoholic beverage. The word *whisky* itself is derived from the Gaelic *uisge beatha* (water of life).

The Act of Union in 1707 combined England, Scotland, and Wales into the United Kingdom. The new UK government in London levied excise taxes on Scottish-made whisky, while at the same time reducing taxes on English gin, leading to a predictable boom in illicit distilling in Scotland. The Excise Act of 1823 both reduced taxes on whisky and coincided with the dawn of the Industrial Revolution, when entrepreneurs began to build legal distilleries on a large scale. Because of the cold northern climate, wood was precious in Scotland, and the decomposing layers of plants in bogs known as peat became the standard household and industrial heating fuel. This same peat was burned to dry the malted grain in distillery kilns, giving the spirit a distinctive smoky flavor.

A more efficient continuous (or column) distillation process was invented by the Irish engineer Aeneas Coffey during the Industrial Revolution in the 1830s (see Chapter 2). These two-column stills separate out the different products of distillation in tall metal columns that can be operated continuously, unlike the older pot stills that need to be cleaned out and recharged after each distillation. Column stills became widespread in the 1860s, and whisky-makers began blending the high-volume whiskies they produced from cheaper unmalted barley with traditional malted Scotches. The blandness of the grain whisky from the column stills toned down the assertive smoky, peaty character of the pot-stilled malted whiskies, producing a smoother beverage. The English embraced the change, turning to Scotches in the 1870s when the insect vine-root pest phylloxera disrupted supplies of wines and brandies from Europe. Since that time, blending companies have purchased many malt distilleries, so that the market is now dominated by huge volume brands that blend flavorful malt whiskies from pot stills with blander grain whiskies from column stills.

Whisky appeals to those looking for an evolved spirit with complex aromas and flavors, a long and interesting story, and a sense of place. Scotland's long human history, sea mists, deep lochs with monster lore, complex geology, heather-clad hillsides, peaty moorlands, and seaweed-fringed islands all contribute to the character of Scotch whisky. A particular influence is heather, a collective name for various species of heath-loving herbs that love poor, acidic soils, and, to deter insects, produce complex chemicals that make their way into the surrounding water and air.

The flavors of whisky are influenced as well by the kinds of rocks that line

Scottish waterways. Scotland has very complex geology because of continental drift. Hundreds of millions of years ago it was part of the North American plate, but it joined the European plate during a huge geological collision: the dividing line lies almost under Hadrian's Wall. The oldest Scottish rocks were thus formed 600 to 800 million years ago, and the water they supply to Islay distilleries gives iron-like flavors. By contrast, granite dominates in the Highlands, and this hard rock doesn't yield much to the water, leaving it soft. The River Spey, although it arises in the Highland granites, is influenced by the substantial limestones and sandstones in its catchment.

Whisky-making is complex. It involves a specialized vocabulary for the mixture of ingredients at different stages, as well as for the vessels used for fermenting, distilling, and aging. The process we describe here is an abbreviated version; there are countless variations on it both in Scotland and elsewhere in the world.

Barley grain contains mostly starch, which is a complex sugar. In order to obtain the component sugar, maltose, that is used for whisky, the grain is steeped in water to let it germinate on a malting floor. Germination takes about five days, and the grains must be turned regularly during this time so that the germination is even. Only a few distilleries, such as Balvenie and Laphroaig, still produce their own maltings. Once the germ, or sprout, is two-thirds the length of the grain the starch has turned to maltose, and the germination process is halted by spreading the wet barley on grids in the kiln and drying it out to 4 percent moisture. Adding peat to the kiln fire gives the whisky a strong smoky flavor. Steam from the kiln fire escapes through the very un-Scottish pagoda-like roofs of the distilleries.

In Scotland, Australia, Japan, and occasionally Ireland (but never in the United States or Canada), peat may be used to various degrees to dry the malt in the kiln fire, giving the malt a distinctive smoky flavor. Some Islay distilleries use malt that has been heavily peated, with phenol levels (by which smoke is measured) as high as fifty parts per million or more; the Speyside average is two to three parts per million. Big-peated whiskies evoke whiskies of the past, instilling authentic flavors that hint of candles and open fires, and different peat sources can give different flavors.

More than a hundred years ago, the Lowland and Speyside regions of Scotland converted to coke and coal fires that burn more cleanly, and don't impart a smoky flavor profile to the drying malt. With few alternatives, however, remote

Highland and island distilleries (for example, those on the islands of Islay and Orkney) have had no choice but to continue to use local peat. Since peat is a non-renewable and very precious energy source, one that is produced very slowly by natural processes of plant decomposition at low temperatures, some distilleries have developed ways to reduce their use of it: for example, by sprinkling peat dust on a normal fire to create the desired smoke during the usual smoking time of eighteen hours, or by running the peat smoke over the malt several times to create the preferred level of smokiness.

The finished dried malt is milled to flour, and the resulting coarse powder, called grist, is mixed with hot water in the mash tun. Depending on the distillery and region, sparging (sprinkling hot water over the grain bed to rinse off the sugars) is often undertaken at two or three different temperatures: a first run at 65 degrees C, a second run at 80 degrees C, and a third run at 95 degrees C, or nearly boiling. Next, the malt is mashed three times before the resulting sugary solution, known as wort, is cooled and pumped into large tubs called washbacks, where yeast is added. During the fermentation process, which takes a few days, the yeast turns the sugars into alcohol and carbon dioxide. Washback tubs are generally covered to prevent unwanted yeast strains entering, and have a blade that continually cuts the foam on the surface of the fermenting wort. These tubs were long made from Oregon pine, which is resistant to fungal attack, but stainless steel is the material of choice nowadays, particularly for New World whiskies. The beer or "wash" resulting from fermentation has an alcohol content of 8 or 9 percent and is ready to go to the still.

It is difficult to believe, but the process of distillation is such a complex fusion of science and art that even the shape of the still influences the flavor of the resulting whisky. A long slim shape produces a softer, purer spirit (like Glenmorangie), while a shorter, squat shape produces more intense flavors (such as a Lagavulin). Distillation usually takes four to eight hours and is repeated at least twice, once in a wash still, and then in a spirit still. The product of the wash still is called "low wine," and has an alcohol concentration of around 20 percent. Spirit stills must be replaced every fifteen to twenty-five years, when the inside walls have given up so much copper to the boiling spirit that the wall thickness has decreased to only four or five millimeters (Figure 12.1).

Distillation of alcoholic beverages relies on the different boiling points of the different components of the beer or wash, as explained in Chapter 2. This difference gave humans a fairly simple way of separating the alcohol from the

Figure 12.1. Row of copper pot stills at Scotland's Lagavulin Distillery.

water—just heat it up, but very carefully. As the low wine is heated just past the boiling point of ethanol (78.8 to 79.4 degrees C), the unpleasant chemicals called "heads" (a.k.a. "fores") begin to vaporize. Heads are a mixture of some ethanol, poisonous methanol, and a witch's brew of other unpleasant chemicals such as acetone and aldehydes that smell and taste like solvents and nail polish remover.

The skill of distillation is to remove the heads, and to focus on the next group of chemicals to vaporize between 82 and 94 degrees C. These are mostly ethanol, some water, and pleasant flavor attributes such as phenols and guaiacols: the "heart" of the distillation. Finally, as the low wine continues to be heated and the temperature exceeds about 95 degrees C, the "tail" of the distillation begins to boil. The tails contain unpleasant-smelling and bitter-tasting compounds such as fusel oils, propanol, isopropanol, and esters. These are not poisonous, but they are generally unwanted (although a small proportion may be retained for more strongly flavored whiskies such as heavily peated Islay malts). Together the heads and tails of the distillation are called feints, and they are often recycled into a future distillation to extract remaining ethanol and positive flavor elements.

Each of the three phases of distillation—the heads, hearts, and tails—bleeds into the next, so that there are "late" heads compounds in the beginning of the heart, and similarly significant "early" tails at the end of the hearts. The flavor profile of a newly made whisky is made up of the delicate balance of the hearts, plus late heads and early tails. Thousands of chemical compounds are involved in the distillation process, and each can add to or subtract from the final flavor. The art of the distiller is to know when to stop and start the heart so as to include the desired compounds and nothing else. Usually the heads take about thirty minutes to run through, hearts three hours, and tails the remainder. Because of the formation of a temporary fusion (azeotrope) between water and ethanol, the purest that ethanol can be distilled to is 96.5 percent ABV. The newly made whisky is usually diluted with water to about 65 to 70 ABV and then transferred to an oak barrel.

Originally oak casks were simply a vessel for transport, but consumers soon realized that the longer the whisky spent in oak, the smoother its taste. Former sherry casks (made from the pedunculate oak, *Quercus robur,* from the Atlantic coast of northwest Spain) were once used to age Scotch whiskies. But over the decades, as the sherry market declined, fewer casks were available. What's more, after the death of the Spanish dictator Francisco Franco in 1975, sherry was shipped out of Spain not in casks but in bottles. As a result, Scotch whisky-makers now frequently employ used bourbon casks from the United States, creating a feedback loop between the American and Scottish whisk(e)y industries.

The oak used in the United States for bourbon whiskey barrels is the local *Quercus alba,* and by law bourbon-makers must use new, charred-oak barrels for aging their whiskeys. The wood is heated to 200 degrees C for thirty minutes to char the inside of the oak barrel, during which time some caramelization of sugars in the wood occurs, lignins are partially converted into vanillins, and other changes happen (see Chapter 6). The charcoal layer on the inside of the barrel is important for removing the harsh flavor components of newly made bourbon, mellowing the spirit. The American Standard Barrel (ASB) holds two hundred liters, and after initial use these are broken down into their individual staves and remade into Scottish hogsheads with a slightly larger capacity of 250 liters. In Portugal, European oak is made into large casks (called butts or pipes) holding five hundred or six hundred liters, which are ideal for the maturation of sherry and port, and later for maturing or finishing Scotch whiskies.

The casks account for 10 to 20 percent of the cost of Scotch production, so they are precious and reused three to four times. (Reuse is not possible in Australia, where the barrels are exhausted after a single use.) Casks can be scraped out and retoasted between fills, enhancing the flavors they impart. A bourbon cask, used from its new state for two to four years in the United States, can be used next in Scotland for thirty years. During the first five to eight years in cask, Scotch whisky loses the harsh flavors of the raw spirit in a process called "subtractive maturation," then begins to take on characteristics of the oak, vanillin, toffee, and so forth in the "additive" maturation phase. During a relatively short final "finishing" phase, many distilleries place the nearly finished Scotch in casks that previously held table wine, port, or sherry; these also have a significant impact on the flavor of the finished product.

The law requires that each cask in the warehouse or bond store be uniquely marked with the name of the distillery and distillation year. This is because tax has not yet been paid on the spirit at this stage, perhaps a decade away from sale, and the excise officer wants to be sure that no whisky goes missing during maturation. The temperature and humidity of the warehouse influence the maturation of the whisky in cask. The Scottish islands have a mild climate due to the influence of the Gulf Stream, while the Highlands have comparatively warmer summers and colder, snowy winters. The alcohol content of the whisky in barrel usually decreases by 0.2 to 0.6 percent annually, and the fluid level decreases by 1 to 2 percent each year in Scotland, 4 to 7 percent in Australia, and as much as 12 percent in warmer production areas such as India. As mentioned earlier, this lost volume is called the "angels' share," and, unlike wine casks, whisky casks are not topped off. After ten years or so, a Scottish cask may have lost fifty to sixty liters to evaporation, or about 20 percent of its volume, and the alcohol content will have declined, from say 63.5 percent to around 58 percent ABV. Bottled whisky must contain at least 40 percent ABV, a ratio achieved by adding water; but it is sometimes bottled at cask strength, which can be as high as 70 percent or so.

Aging in cask does a number of things to the flavor and color of a whisky, and probably accounts for up to 70 percent of the flavor of the finished spirit. Wood expands and contracts through the seasons and from day to night, and the whisky penetrates the wood accordingly. Some of the atmosphere of the barrel house may enter the cask. Sherry and bourbon barrels will impart elements of the liquid they once contained. If the cask has been charred, the whisky penetrates a layer of charcoal, further filtering it and removing harsh elements such as sulfur notes (which smell like a struck match or rubber). Whisky in large casks must be stored longer than whisky in small casks, because of the different internal surface area of the casks, and European oak tends to be more porous than American, allowing for more oxidization.

One of the most important reactions is the slow oxidation of the whisky in barrel over many years. Oxygen in the headspace dissolves into the spirit and begins numerous chemical reactions that increase the complexity of agreeable flavors in the whisky, especially fragrant, fruity, and spicy notes. Traces of copper from the still are a catalyst for a series of chemical reactions in the whisky. They convert oxygen to hydrogen peroxide, which attacks the wood and releases vanillin. The brown color of whisky comes from wood components that are soluble in alcohol, including tannins.

Some distilleries use chill filtration after maturation, to stop the final whisky from going cloudy if the consumer adds ice and thereby causes residual proteins and other elements to form a haze. But most tasters can't tell the difference between chill filtered and not—and shouldn't add ice to their whisky anyway. Moreover, filtration removes some flavor elements, such as fatty acids, of which the presence or absence is a matter of personal preference.

Only a few distilleries have their own bottling plants (Glenfiddich, Springbank, Bruichladdich); the other distilleries ship their finished casks to the big plants in Glasgow, Edinburgh, and Perth.

Scotland has about a hundred distilleries, of which 80 to 90 percent are operating at any one time. There are many disused, abandoned, and dismantled distilleries in Scotland, attesting to the boom-and-bust nature of the Scotch industry.

Whisky labeling is extremely complex, much like Scottish geography, the Scottish character, and even the spirit itself, but here is a simplified guide. The only things that can be in Scotch whisky are the spirit, water, and limited added caramel coloring.

Single Malt Scotch is made from barley malt produced in a single Scottish distillery, and is usually more or less peated. It may be a mix of whiskies from different years, and it is made by double-distilling in pot stills (column stills were outlawed for single malt Scotch in 2009). Because the blend may be of different ages, any age statement on the bottle must refer to the age of the youngest spirit in the mix. Aging is done in used wooden bourbon or sherry barrels, for a minimum of three years. A controversy raged in 2003 when Cardhu single malt was relaunched as pure malt after the label switched its product from a single malt to a blended malt Scotch (vatted malt).

Blended Malt Scotch (Vatted Malt) is made from a blend of malts from different distilleries. "Blended malt" and "pure malt" mean the same thing as vatted malt. A proposal by the Scotch Whisky Association (SWA) to call these blended malt Scotch whiskies created confusion because the word *blended* is used to indicate an initial blend of grain and malt, not a mixture or vatting of malt Scotches from different distilleries.

Scotch Grain Whisky is made from wheat or corn, with a small percentage of unmalted and malted barley. It is made in column stills and must be matured for a minimum of three years in oak.

Blended Scotch Whisky is a blend of grain and malt whiskies. The percentage of each varies, and they can include many malts from different distilleries. It is usually a blend of highly flavored island or Highland whiskies, with more neutral spirit from the lowlands. The vast bulk of Scotch sold is blended in brands such as Johnnie Walker, Dewar's, and Chivas Regal.

Irish Whiskey is a blend of malt and grain whiskeys. The Irish generally do not use peat, or only a little of it. Irish whiskey is distilled three times, in both column and pot stills. It is aged in used bourbon or sherry casks for a minimum of three years. Note that in Ireland and the United States a different spelling for oak-aged grain spirit is generally used, with an "e" inserted between the "k" and the "y" (see Chapter 3). Well-known Irish whiskey brands include Bushmills, Tullamore, and Jameson. Boutique distilleries are proliferating in both Scotland and Ireland, and the distinctions between them are now blurred.

The Scotch Whisky Association, or SWA, champions the cause of Scottish distillers, blenders, and others involved in Scotch production. Among other things, the SWA advances rules for Scotch production and controls the names of the regions that can be applied on Scotch labels. In 2019 the SWA produced the latest version of the Technical File that verifies what is, and what is not, Scotch whisky. It also provided definitions of the five different geographic regions (geographic indications, or GIs) for Scotch whisky, much as is done in quality wine regions around the world. Although some regional differences in flavor are apparent, it is often impossible to associate a particular whisky with a region based on taste alone. The Highlands is the largest of the regions, and includes all the islands except Islay. Some people separate islands such as Orkney, the Hebrides, Skye, Mull, Jura, and Arran (but not Islay, which is its own region) into their own subregion of the Highlands. The Lowlands region extends from the southern border of Scotland, north to a line running between the Tay Estuary on the east coast to the Clyde Estuary on the west coast. There are three other relatively small regions: the island of Islay with its deeply flavored peaty malts; Campbeltown on the tip of the Kintyre Peninsula, which has its own spicy, salty regional style; and Speyside, a region along the valley of the River Spey.

Lowlands: This region has only three distilleries in operation (Glenkinchie, Auchentoshan—the last to triple distill—and Bladnoch), which make accessible, light, grassy, herbal, mild-bodied Scotch.

Highlands: Many distilleries (for example, Glenmorangie, Dalwhinnie, and Ardmore) lie in this broad geographic area, with its complex terrain, geology, and vegetation. Highland whiskies tend to be light, fruity, and spicy.

Islay: This island has nine distilleries (including Bruichladdich, Laphroaig, and Ardberg) known for the distinctively smoky, medicinal, peaty taste of their product.

Speyside: This area has many distilleries (perhaps 50–65 percent of all Scotch distilleries), among them The Macallan, Glenfiddich, and Glenfarclas. They make Scotch with complex, sweet, and rich flavors.

Campbeltown: This area has similar smoky characteristics to Islay, with three distilleries (Glengyle, Glen Scotia, and Springbank). Whisky production was a much more prominent industry here 250 years ago—in 1759 there were no fewer than thirty-two distilleries in the area.

Of all the spirits, whisky has traveled most around the world, achieving a geographic and stylistic diversity exceeding all others. Indeed, Scotland may have adopted whisky distillation from Ireland (although documentation for distilling in Scotland goes back to 1495, significantly earlier than equivalent evidence in Ireland), and the whisky industry in the United States probably has Irish roots as well. All other regions have budded from the basic Scottish tradition.

Initially, the Irish monasteries distilled whisky for medicinal purposes, as did the monasteries they established in Scotland, just across the North Channel. Some think the first distillates may have been fruit-based rather than grain-based. For its part, barley-based Irish whiskey first appears in the historical records in the mid-sixteenth century, and early Irish whiskeys may have been flavored with raisins and herbs such as fennel seeds. Queen Elizabeth I was said to have been fond of Irish whiskey, and regularly had casks shipped to London. An excise tax imposed in 1661 created an illegal Irish whisky industry (*poitin,* Irish version of moonshine), but large legal distilleries still operated in Dublin and other major commercial centers such as Cork and Galway. By the end of the eighteenth century, two thousand legal stills were operating in Ireland, and under English rule Irish whisky was exported all over the expanding British empire. During the late nineteenth century, more than four hundred brands were being exported to the United States.

When Prohibition was established in the United States, the largest market for many Irish whiskeys declined and many distilleries went broke. The Great Depression and World War II posed further challenges for the Irish whiskey distilleries. In 1966 the three remaining distilleries in Ireland (Powers, Jameson, and Cork) merged into the Irish Distillers Company (IDC). In 1972 Bushmills in Northern Ireland also joined IDC, which in 1975 opened a huge distillery near Midleton, County Cork and closed the other distilleries in the Irish Republic. Irish whiskey tends to be fruity, light, and less peaty than Scotch.

During the early eighteenth century, difficult economic times and religious upheavals led many Scottish and Irish immigrants to make their way to North America and elsewhere, taking with them their skills and expertise in distilling spirits. Many of the American immigrants settled initially in Pennsylvania, Maryland, and Western Virginia. Poor roads meant it was difficult to transport farm produce to major markets on the East Coast, so the farmers began distilling their excess grain crops into whiskey for sale locally, and for as far as the barrel could travel. Pennsylvania developed a primarily rye whiskey, whereas farther west and south, corn whiskeys were prevalent.

By the end of the American War of Independence in 1784, the first commercial distilleries were already operating in western Virginia (then part of Kentucky). In 1794, when the federal government levied an excise tax on distillers, western Pennsylvanian distillers reacted strongly. So began the Whiskey Rebellion, an episode so violent that some tax collectors were killed (see Chapters 3 and 20). Following this unrest, a new group of settlers from the Cumberland Gap to Kentucky and Tennessee began distilling after finding good land for growing corn and smooth, limestone-filtered water.

The word *bourbon* probably derives from a county in eastern Kentucky that was a center of whiskey production in the early nineteenth century but is ironically dry today. Alternatively, some believe the name comes from Bourbon Street in New Orleans, where much Kentucky whiskey was consumed. Either way, bourbon producers age their spirit in new charred American oak casks and use a "sour mash" technique. This process, designed to maintain consistency, involves adding a small percentage of the previous fermentation to the washback as a starter for the next one. By the 1840s bourbon was being marketed as a distinctive American style of whiskey and produced in a wide range of eastern states, although the only legal geographic requirement is that it be made in the United States. Initially bourbon was made in pot stills, a process that is making a comeback in some craft distilleries, but most is now made in continuous stills.

The U.S. temperance movement gained steam in the late nineteenth century in various counties and states, and national Prohibition was introduced from 1919 to 1933. This played havoc with the industry because it interrupted production and aging. Whiskey consumption continued, as drinkers turned to illegal moonshine and Canadian products of dubious quality, but Americans' tastes in whiskey changed, because the products available were lighter in taste and body than their predecessors had been. By the closing decades of the twentieth century there were only ten distilleries left in Kentucky, and two in Tennessee (including Jack Daniels).

Nowadays an American bourbon whiskey must have a minimum of 51 percent corn in its "mash bill," or mix of grains. It must be produced in the United States, distilled at less than 80 percent ABV, and aged for a minimum of two years in barrels made of new charred American oak (*Quercus alba*). There can be no flavorings or coloring agents added, otherwise the result is a "blended" whiskey. Tennessee whiskey is almost identical to that produced in Kentucky, although Tennesseans filter the product through sugar-maple charcoal (in what is called the Lincoln County Process) for a smooth, mellow flavor. Generally, American whiskey is sweeter, less smoky, and less peaty than its Scotch counterpart. Rye whiskeys (which have a minimum 51 percent rye mash bill) continue to be made in Pennsylvania and Maryland, but their stronger, harder flavors meant that they suffered more after Prohibition, and production has almost entirely died out. Their dry, peppery, astringent flavors require at least four years of barrel aging to soften.

Blended American whiskeys began to be produced in the early nineteenth century, after column stills allowed a lot of neutral spirit to be produced quickly. Distillers blended bourbon and rye whiskeys with neutral spirits to create their own blend and to stretch their supply of straight whiskeys. The result, which must contain a minimum of 20 percent straight whiskey, generally tastes bland in comparison to bourbons; even so, they enjoyed a sales boost after World War II.

Corn whiskeys are unaged spirits that are considered precursors to bourbon. When excise taxes were introduced during the Civil War, much of the corn whiskey produced went straight to moonshine, and stayed there. To be called corn whiskey, the mash bill must be a minimum 80 percent corn, distilled to a maximum 80 percent ABV. Corn whiskey is exceptional because it doesn't need to be aged to reach full flavor potential, even though sometimes it spends a short period in barrel. A variant called straight corn whiskey is aged for two years or more in new or used uncharred barrels.

Canadian whisky maintains the Scottish spelling because, unlike the waves of Irish immigrants that came to the United States during the early 1800s, most immigrants to Canada during this period were from Scotland. Many early Canadian distillers began as millers, who distilled excess rye and wheat starting in the 1830s. The Canadian industry has always been strongly influenced by the political and cultural changes south of the border. Canadian whisky found a large market in the United States that expanded during the Civil War, although Prohibition later caused many legal Canadian distilleries to go under. After Prohibition, the diminished Canadian whisky industry developed its own style, quite unlike American whiskey, and has since suffered because of the different definitions of "blended" whisk(e)ys in the United States and Canada. Nonetheless, despite the confusion over labeling, Canadian whisky was popular in the United States from the beginning; in fact, from the Civil War until the early twenty-first century, more of it was sold in the United States than bourbon.

In the United States, blended whiskey needs only to contain a minimum 20 percent straight whiskey; the remainder can be neutral alcohol, flavorings, and caramel. This means that blended whiskeys have an inferior reputation in the marketplace, and are generally less expensive than bourbons. Canadian whisky almost always uses a strong element of rye in its mix, which adds spicy, peppery notes; but it also uses distillates from milder corn, wheat, and barley grains. In a significant departure from American producers, Canadian distilleries make spirit from each grain separately, then blend them together. This technique gives Canadian producers tremendous control over the flavor of the resulting whisky, depending on the proportions used of spirits from different grains. Canada even allows the addition of up to 9 percent of another finished liquor, such as sherry. Canadian whiskies are aged primarily in used oak barrels for a minimum of three years, in much the same way as Scotch whiskies are.

The origin of Japanese whisky can be traced to the son of a sake producer, Masataka Taketsuru, who visited Scotland in 1918. He spent two years studying chemistry at Glasgow University and working in the Highland distilleries Hazelburn and Longmorn before returning to Japan in 1920, with a Scottish bride and a determination to change Japanese distilling. In the 1920s the Japanese were major consumers of Scotch. In 1923 Shinjiro Torii built a whisky distillery called Kotobukiya in Yamazaki, Kyoto, an area famous for its pure water, and hired Taketsuru to produce barley malt and grain whiskies based on Scottish recipes and even some made with Scottish peat, which became very popular locally. Torii later went on to found Suntory, and Taketsuru founded his own distillery, Nikka, in 1934. While Japanese whisky began with Scottish recipes, it has evolved its own style, focusing on purity of flavor. For maturation, the aromatic species of Japanese oak called Mizunara (*Quercus mongolica*) is used (see Chapter 6).

Whisky production and consumption have a long history in Australia. Australia was Scotland's largest whisky export market until the late 1930s, and whisky has been distilled in Australia since 1791, only three years after the first settlers arrived in Sydney. Victoria became the main whisky-producing state from the mid-nineteenth to mid-twentieth centuries, with Federal Distilleries in Melbourne producing about four million liters a year in the late nineteenth century. Some Scottish and English distilling companies built Australian distilleries in the early twentieth century, aiming to produce cheap young whiskies that could compete with the more expensive imported Scotch. When import tariffs were removed in the 1960s, the local industry collapsed, unable to compete with suddenly less expensive but higher-quality Scottish imports.

The modern Australian craft whisky industry consists of small-scale distilleries scattered across the southern half of the continent, from Limeburners in the far southwest of Western Australia to Castle Glen in southeast Queensland. The epicenter of the Australian whisky industry, however, is based offshore, in Tasmania.

The Tasmanians Bill and Lyn Lark are credited with beginning the modern Australian whisky-producing era in the 1990s, when they lobbied the government to reduce the minimum legal size of stills from 2,700 liters to something more manageable for artisanal producers. Australian producers are now in an expansive experimental phase, producing whiskies in both the Scottish and Irish styles, using various grains and their own local peat, which adds an Antipodean blend of flavors to the maltings. Lark and other early producers like Hellyers Road began by purchasing their wash from local beer breweries, and malted barley is now purchased locally from a Tasmanian malting house. Aging occurs in old bourbon and Spanish sherry casks, or from casks sourced from Australia's extensive fortified wine industry. In fact, the demand for used fortified barrels for whisky maturation is outstripping supply, because whisky has become more popular than fortified wine.

The Tasmanian whisky industry is currently riding a wave of unparalleled success, with booming domestic sales and exports driven by critical acclaim. Among many other awards won by Tasmanian whiskies, Sullivans Cove French Oak Single Cask was awarded world's best single malt in the World Whiskies Awards for 2014. The success of the industry is creating pressure for the early release of stock already aging in bond houses, a good problem to have.

Local producers are sure the Tasmanian climate has something to do with the high quality of Tasmanian whisky. Daily and yearly changes in temperature are very different from those in Scotland. Tasmania is still many degrees of latitude closer to the equator than Scotland, and hence is warmer on average. But cold fronts can sweep up from Antarctica during winter, plunging the state into a deep freeze, and in summer high-pressure systems can bring sweltering air from the deserts of the mainland across the narrow Bass Strait. Indeed, temperatures in Tasmania can climb above 40 degrees Celsius in summer and may fluctuate by double digits over twenty-four hours. These changes are much greater than you would see in a Scottish barrel house, and they mean that the barrels "breathe" much more, and more often. As the barrels expand with the heat, whisky seeps deep into the wood of the barrel, removing harsh attributes and soaking up smooth oak barrel flavors of vanilla and the previous contents, whether bourbon, sherry, or port. Once cooled the barrels shrink, and the liquid is forced back out of the wood. Hence the contents of a Tasmanian barrel may soften and take up as much flavor in five to six years as you would expect those in a barrel of Scotch to do in ten to eighteen years.

The simplest way to consume Scotch is neat (straight), at room temperature—which is how the distiller wants you to taste it. This is not usually how most drinkers start, however, because the strong flavors and high alcohol content (40 percent ABV at least, and more than that for cask strength) can be daunting. Whisky drinkers often add just a few drops of water to "break the spirit," or perhaps ice or soda. But ice can dull your taste buds and dilute the spirit too much, so watch how much you add. Whisky is the spirit base of many cocktails, including the Whiskey Sour, made from bourbon, lemon juice, sugar syrup, bitters, and egg white, and the Manhattan, which includes rye whisky, vermouth, bitters, and a cherry. An Old Fashioned is made from bourbon, orange peel, a sugar cube, bitters, and a dash of soda.

Connoisseurs use various specialized glasses for whisky, but the most relevant characteristics are the shape and size of the bowl above the liquid, which holds and concentrates the aroma. The big, open, heavy glasses that are often sold for whisky are actually not good for this. At Tim's distillery we often use a small wine glass for whisky tasting, and proper whisky tasting glasses have a similar shape. The smell of whisky is very important to its taste, because most of our "taste" is derived from our sense of smell (Chapter 8). Before smelling the whisky, you should examine its color, clarity, and viscosity. The color of the whisky can be a hint as to its age (generally darker is older), and to the kind of casks used. (Be careful, though, because artificial coloring with caramel is allowed for Scotch.) Chill filtration is used to increase the clarity of whisky when it is less than 46 percent ABV, but an expensive, nonfiltered whisky might become cloudy when diluted with water. Finally, the more legs a whisky has, the more alcoholic and/or older it probably is.

Aahh, the smell of whisky! Aromas can come from many sources, such as the type of grain used, the malting process, fermentation and distillation, and of course the kind of wood it was aged in and the history of the cask. Bring your nose close to the top of the glass, and take a long, slow sniff. Don't swirl the glass as if it were wine—that is the sure sign of a beginner. If you like, you can move the glass in your hand to an almost horizontal position (be careful not to spill it) and see if this changes the aromas you detect—it usually does. The aromas you will find range from floral to fruity, malty, spicy, woody, and smoky. Humans are very visual creatures, and our sense of smell is buried deep in our brains (Chapter 8). When we are tasting whiskies, we always try to get into a very mindful space and let the aromatic experience wash over our deep brain without letting external influences intrude.

It can be challenging to describe a whisky aroma without some kind of simile and metaphor, such as "quite spicy and herbal, with smoke and cedar undertones." But how would you explain to somebody the color orange, without saying the word *orange* or pointing to it on a color wheel? Much of the communication about our senses involves learning such conventions. It is the same with whisky, and many of the conventions come from the experience of trying many different whiskies with good friends. This is where Tim's dinosaur taste-shape metaphor comes in handy. Did the whisky enter your mouth smoothly, flow over your tongue with graceful elegance, and pass serenely to the back of your mouth? During that journey did the whisky aroma and taste please your brain in a harmonious way, or was it a confusing trip? Don't be self-conscious; we all know what we like. Most of all, whisky is a drink for sharing, and for conversation. The most important part of drinking whisky is who you drink it with, and where. You can drink an expensive Scotch in a glass balloon at the Vaults in Edinburgh, or you can drink whisky in an enamel mug around a campfire with good friends. Either way, the drink in your hand is the best whisky in the world.

13

Gin (and Genever)

Ian Tattersall

We were intrigued. In front of us in an angular bottle was a gin flavored with botanicals from the most unusual and intensely local flora found anywhere in the world—the Southern Cape fynbos of South Africa—and made in a wood-fired still by vapor infusion using a base of neutral cane spirit. On the nose, a comfortingly familiar note of juniper was quickly supplemented by an intensely floral bouquet unlike any other we could recall. On the palate the floral quality predominated, backed up by some citrusy and herbal tones and a slight hint of licorice, all bound together in an unexpectedly creamy mouthfeel. Surprisingly, adding a strip of lemon zest seemed to flatten all of these sensations a bit. A hint of orange peel might have suited better, but honestly, we weren't sure we wanted to mix this spirit with anything—though in the end a splash of tonic didn't hurt. It seemed to us that this unusual gin's delicate fragrances made it an ideal sipper for anyone who might appreciate the way its layers of both familiar and exotic aromas gradually reveal themselves.

Mother's Ruin. Dutch Courage. Babyshambles. Such nicknames make it clear that one of the most sophisticated spirits available on the market today has not always enjoyed the most luminous of reputations. Yet gin, nowadays broadly defined as a spirit made by distilling a grain mash with juniper berries and other botanicals as flavorings, appears historically to have started life as a medicinal tonic aimed at ailing citizens. An Egyptian papyrus dating from 1550 BCE reportedly prescribes juniper for jaundice, while in the first century CE Pliny the Younger recommended it for a range of symptoms ranging from flatulence to coughing. And it was in a clearly medical context that the Flemish poet Jacob van Maerlant included the earliest known recipe for cooking juniper berries in wine in his gloriously illustrated *Der Naturen Bloeme* (1269). A couple of centuries then passed before the first known instructions for infusing juniper in a distilled base appeared, in *Gebrande Wyn te Maken* (circa 1495), one of a collection of Dutch medical manuscripts that now reside in London's British Library.

Gebrande (burned) *Wyn* is the term from which the English word *brandy* is derived; and the anonymous 1495 recipe for making it prescribes infusing a bewildering (and in those days almost unimaginably expensive) variety of exotic spices such as nutmeg, cinnamon, and cardamom in a wine distillate, along with more humble local ingredients such as sage and the juniper berries that mark this particular concoction as a gin precursor. A century after this hesitant if amazingly extravagant start, grain, rather than grape, had become definitively established in Holland as the standard base for distilled spirits. Those spirits included the *genever* (a.k.a. *jenever*) that was named for the juniper that flavored them. Soon the uses of genever became much more recreational than medical, for while juniper proved a bit of a disappointment as a cure for the plague, the beverage it flavored was a surprisingly effective morale booster. As always, the authorities soon took notice of this potentially lucrative new source of public enjoyment, and the city of Amsterdam started taxing spirits as early as 1497. Despite the added expense, a century later the place was full of distilleries.

One of the main reasons that grain spirits enjoyed early popularity as a recreational drink in the Low Countries was that the region was entering the Little Ice Age, a climatic event that gripped the region well into the nineteenth century. It was simply too cold to grow grapes, so a local wine industry that might have competed with the grain distillers could never really get going. The appeal of grain spirits quickly spread to Great Britain, just across the North Sea. The English, most of whom had been stuck drinking weak and rather uninteresting "small beer" (derived from the second or even a third pass of the mash) as an alternative to their frequently polluted water, discovered the spirit while at war. During the dreadful Thirty Years' War that occupied most of the first half of the seventeenth century, large numbers of British soldiers sought "Dutch courage" on the bloody battlefields of the Netherlands and elsewhere in Europe, and many of the survivors eventually took home with them a newly acquired taste for the genever that supplied it. The novel spirit first established beachheads in the port cities of southern England; aided by its medicinal reputation, it then rapidly caught the fancy of drinkers in London and beyond.

British mangling of the Dutch name resulted in a generation of "ginever" consumers, who soon shortened the name to "gin." The abbreviated term seems to have first appeared in print in 1714, in Bernard Mandeville's *The Fable of the Bees; or, Private Vices, Publick Benefits.* Mandeville, a moralist and philosopher who was actually of Dutch origin, alluded to the beverage as "the infamous liquor." This description seems prescient given what would unfold, but at the time Mandeville's disapproval may have made him appear a bit of a spoilsport.

A quarter-century before Mandeville wrote, the Dutch Prince William of Orange and his wife, Mary (James' Protestant daughter), had jointly ascended to the British throne following the flight to France of the Catholic King James II. William brought to London not only his Dutch bibulous tastes, but also his legendary dislike of the wine-and-brandy-drinking French. Suddenly, genever was the preferred beverage at the Royal Court; and in 1690 Parliament passed an "Act for Encouraging the Distilling of Brandy and Spirits from Corn" that allowed almost anyone in Britain to engage in distilling. The aim of this legislation was to improve the availability of substitutes for the French wine-based products that had been throttled by a blockade and heavy taxation; at the same time, a modest duty on British spirits conveniently helped fund the hostilities with France. Drinking genever/gin accordingly became downright patriotic, and very much the Protestant thing to do.

Given these political and economic developments, it is hardly surprising that by the end of the seventeenth century, gin—both imported and locally produced—had become England's spirit of choice. The malty Dutch genevers typically went to those in the upper economic strata, while at the lower end of the market the leaner, and usually harsher, local product was favored. This was of course an economic preference, because gin could be made not just from untaxed local ingredients, but from cheap and inferior ones too. Most importantly, grain spirit could be inexpensively distilled from barley that was not good enough to use for making beer, and flavoring it with juniper (and sometimes even turpentine) helped disguise its gustatory defects. The cheap—if probably not always cheerful—resulting beverage was often less expensive than beer, which had been hit by heavy taxes in the 1694 Tonnage Act. At the very least, it got you a lot drunker for your money.

Still, gin was not yet an entirely unregulated product because, despite the passage of the 1690 Distilling Act, the Worshipful Company of Distillers had held on to a half-century-old monopoly on primary distillation within twenty-one miles of the center of London. That meant that this one partnership controlled the country's most lucrative spirits market, providing the basic grain distillate to local "compounders" who flavored and distributed the final product. Total deregulation of grain distilling came, oddly enough, only after Mary's brandy-drinking sister Anne had assumed the throne following William's death in 1702. After retracting the Worshipful Company's monopoly, her government soon eased the final remaining restrictions, allowing anyone to distill products for sale after simply posting publicly, for ten days, notice of their intention to do so. As always, there were unintended consequences. Monopolistic as they might have been, the Worshipful folks had maintained relatively high distilling standards for the base spirit, and their loss of control led not only to a huge proliferation in the number of London distilleries, but also to significant variation in the quality of their products.

Gin distillation nonetheless boomed in England during the first half of the eighteenth century, especially in the country's crowded capital, which was full of grindingly poor people anxious to forget their worries. Everyone was in on the act: it's been estimated that in addition to all the commercial distilleries, as many as 1,500 domestic stills may have contributed to the local flow of gin. As a result, gin shops (bars and retail operations alike) proliferated, inaugurating what came to be known as the Gin Craze, which lasted from around 1720 to 1751.

Although official "wars" on anything in society usually go horribly wrong, you can hardly blame the government for trying to put the kibosh on the gin party. The figures are a bit hazy, but even now they seem alarming. It's estimated that in 1700 the average English drinker imbibed a mere quart and a half of gin each year. By 1720, when the Gin Craze officially began, consumption had already doubled to three quarts among what magistrates were wont to call "the inferior sort of people"; by nine years later, the corresponding figure was six quarts or so, rising to as perhaps as many as ten by 1743. Ten quarts a year (actually, at the lower end of the available estimates) is less than one quart a month, which doesn't sound like much (it works out to something like a single one-ounce shot a day, per head of the drinking population); but whatever the actual consumption, it was apparently enough to send things truly out of control among the less affluent classes. It didn't help that during this period some distillers also took to adding a little sulfuric acid to the grain base. The acid did not itself distill, but it did add diethyl ether aromas to the distillate. This made for a slightly sweeter and perhaps more drinkable beverage and, by some accounts, amplified its intoxicating effects.

Starting in 1729, Parliament, in a series of increasingly desperate attempts to stem the tide of drunkenness, passed a succession of Gin Acts aimed at reducing gin consumption and regulating the activities of the gin shops. These efforts mostly drove distillers underground, and retailers into the dankest back alleys. The Third Gin Act of 1736 was especially harsh, imposing particularly high taxes on retailers and consumers and leading to enormous riots. In response the government—which was rapidly running out of money to pay the informers it had depended on for Gin Law enforcement—lowered spirits duties again in 1743. This move made enforcement easier, and gin consumption actually began to drop. But not enough; in 1747 Parliament felt obliged to tinker again. It modestly increased duties on spirits while lowering taxes on beer, changing the economic balance between the two beverages. Finally, in 1751, the Eighth Gin Act raised taxes on locally produced spirits, while at the same time raising the minimum rent for premises licensed to purvey them. It seems to have been the rent angle more than anything else that choked off the backstreet purveyors.

An enduring icon of the era is a pair of engravings that William Hogarth published in 1751, the last official year of the Gin Craze (Figure 13.1). Both prints were intended as pure propaganda for a moral crusade against the spirit. Each image is set in a proletarian area of London; but one, *Beer Street,* is populated by

Figure 13.1. William Hogarth's 1751 prints *Beer Street* (*left*) and *Gin Lane.*

neatly dressed and prosperous workers who are enjoying foaming mugs of Britain's traditional plebeian beverage as a reward for their hard and dedicated labors. The only sign of distress in this scene is a boarded-up pawnbroker's shop, where the impecunious proprietor receives his beer through a hole in the door because he fears the bailiff. "Here," wrote Hogarth, "Industry and jollity go hand in hand."

Beer Street's sibling print, *Gin Lane,* could hardly be more different. This print shows London's poor in the grip of desperate addiction to the foul foreign spirit. A disheveled mother, legs covered in sores, heedlessly allows her baby to tumble over a banister, presumably to its death. To her left a ghastly skeletal soldier is apparently dying, grasping a basket containing an empty booze bottle and a copy of a poem titled "The Downfall of Mrs Gin." Behind these two figures, wretches frantic for a drink pawn their most precious possessions, a man competes with a dog for a dry bone, orphan girls hover around gin barrels, and the local undertaker is seen hard at work among the brawling masses. The message is impossible to miss. As Hogarth put it, "that invigorating liquor [beer] is recommended, in order to drive [gin] out of business." Beer, the quintessentially English tipple, is seen as synonymous with prosperity and social order, while the vile Netherlandish spirit is associated with every possible social ill, including poverty, violence, disease, starvation, and moral decay.

The twin measures of 1751 eventually put paid to the backstreet gin shops, leaving the business in the hands of more reputable retailers. Other factors also helped put a definitive end to the Gin Craze. Probably most important, a series of poor harvests raised the price of the cereals from which gin was made, eventually prompting the government to mount a temporary ban on distilling from domestic grain in 1757—a ban that was not lifted until 1760, following an abundant harvest. Meanwhile, economic recession had reduced disposable incomes among the gin-drinking classes, and technological improvements in the production of the more virtuous beer had made ale a more competitive product in the marketplace. Everything combined to put the writing on the wall for English gin. In 1794 there were still around forty malt distillers, compounders, and rectifiers active in London. But by this point consumers were becoming more discriminating, and local compounders were beginning to turn to grain spirits originating in Scotland, which remains to this day the epicenter of large-scale British gin production.

Traditional Dutch genevers were produced by making a highly extracted malt wine from a mixed bill of wheat, rye, and barley, added in that order to the mash (see Chapter 4). Fermentation, commonly using baker's yeast, followed after the mash had cooled from the initial boil. Next came pot distillation of the resulting "low wine" (Chapter 12). Typically three passes were made, yielding a maximum ABV of around 70 percent. The result was then "rectified" through further distillation with added juniper and other botanicals. This was the basic procedure imported into England in the early eighteenth century; but, as we've seen, during the years of the Gin Craze, British gin manufacture was accompanied by manufacturing shortcuts and a deterioration in the quality of the base ingredients. During the second half of the eighteenth century, more refined English tastes swung back toward the more expensive Dutch product, while the local version went to the impecunious working classes.

Two factors coincided to change this situation, at least to some degree. The first had to do with gin's price relative to beer. In the early 1820s Parliament decided to encourage domestic distilling of gin by drastically reducing duties again; and, between 1825 and 1826, production doubled. Now gin was once again cheaper than beer, creating a new demand for places in which to drink it. Enter the Gin Palace, an establishment designed to compete with the newly licensed ale-selling public houses by offering patrons lavish ambience as well as cheap booze. The first gin palace was opened in 1828 in London's Holborn (not far, as it happens, from the Saint Giles slum that provided the setting for Hogarth's *Gin Lane*), its brightly gas-lit interior featuring a long bar and plenty of mirrors, polished brass-work, and mahogany paneling. It was the first of many such elaborate drinking destinations: within a decade there were some five thousand gin palaces in London alone, each one more grandiosely assuming the same basic role that the grim gin shops had filled eighty years earlier. Belatedly the government recognized its mistake, and lifted the tax on beer. A price-sensitive clientele rushed back to the pubs, and by midcentury the gin palace was already a fading memory.

This all happened because British gin was still a proletarian drink that basically competed with beer. But there were efforts afoot to improve the taste and experience of gin. Since about 1819, the Rectifiers' Club had been meeting regularly to promote better distilling practices, with the particular aim of improving the quality of the base spirits used in gin production. And these goals would shortly be realized by technological changes that would radically transform the production of gin. Starting in 1828, the very year the original gin palace opened, the first continuous still was installed at the Cameron Bridge Grain Distillery in Fife, Scotland. This particular initiative ran into technical and financial difficulties, but a mere two years later it was followed by the patenting of Aeneas Coffey's two-column continuous still, a device that became widely adopted by distillers even as it was shunned by whiskey-makers in Coffey's native Ireland. The neutral grain alcohol of around 96 percent ABV that continuous stills produced was the ideal spirit base for gin production.

It is hard to overstate the importance of this technological leap for the history of gin. For the first time, a gin-maker could use botanicals to create a desired flavor profile, and not just to disguise the rough edges of the base spirit. The resulting product could thus be lighter, more complex, more balanced, and a more accurate transcription of its maker's intentions. Pretty soon, continuous distillation routinely provided the initial grain spirit for gins, with copper alembics still typically being used for rectification.

From the 1830s onward, then, the continuous still presented gin-makers with a relatively blank canvas on which to practice their art of flavoring neutral distillates. Those distillates continued to be primarily sourced from grain, although today pretty much anything is allowed, including neutral spirits derived from beet molasses, potatoes, and even grapes: each resulting spirit, while nearly neutral, still has its own characteristics. The most common method of extracting the flavors in the botanicals, in the early days as now, is known as "steep and boil." Usually, this procedure involves diluting the spirit base to 50 percent ABV, and soaking the botanical mix in the result. Steeping can begin as much as a couple of days before final distillation, or the botanicals can be added just before the pot still is closed. As described in Chapter 5, the mix is then heated to a temperature above the boiling point of ethanol and below that of water, causing both the ethanol and the desired botanical extracts to vaporize and rise through a condenser, eventually emerging as a liquid. Fearing impurities, distillers routinely discard both the liquid "heads" that emerge first, and the "tails" that emerge last—although the gin-maker, starting with a high-alcohol neutral spirit, has fewer worries about unwanted compounds than the makers of, say, whiskies.

An alternative to direct steeping involves "vapor infusion." In this technique the botanicals are placed in a container (often with separate layers for separate botanicals) that is suspended above the grain alcohol in the boiler in such a manner that the hot vapors rising from the alcohol pass through it before entering the condenser. Steepers say that this method fails to extract all the aromas offered by the botanicals; vaporizers reply that they get a lighter, fresher-tasting result. Sometimes a gin distiller will try to get the best of both worlds by combining steeping and vapor treatments, subjecting some of the botanicals to soaking while passing vapor through the others. Indeed, almost anything is possible, and occasionally the botanicals will be steeped and boiled or vaporized individually, then combined. Every option involves tradeoffs; in this last case, some argue that key interactions among the molecules given off by the botanicals are lost, while proponents claim that purity of flavor is better preserved. There is a similar controversy over vacuum distillation, a technique limited to small-batch production. Here the botanicals are steeped and boiled in a vacuum, so that the alcohol evaporates at a lower temperature. Advocates are adamant that this produces more vibrant flavors; but in the end the variables are so numerous in the production of any gin that it is hard for any outside observer to judge.

One process that generally distinguishes craft distillers from major manufacturers of gin is that the small producers tend to use a "single-shot" approach to rectification, implementing the exact recipe for each gin they make. Large-scale producers, on the other hand, often increase the quantities of botanicals relative to spirit during the infusion, then add more spirit to dilute the result to the required proportions. This "multi-shot" approach economizes on valuable still time, and is said by those who use it to allow for more precise implementation of each gin's particular botanical bill. The craft folk, of course, disagree. Once more, the only proof, as it were, is on the individual palate. Gin-making and gin-tasting are bewilderingly complex, although when the distilling is complete, that's normally it. Most gins are bottled right away, and only a tiny minority are ever aged (see Chapter 6). The color you often see in gins might occasionally be a result of barreling, but it's much more likely to come from the botanicals.

When the purity of the base spirit can be taken pretty much for granted, the key to your enjoyment of any gin will lie strictly in the botanicals, the plant essences with which it is flavored. Hundreds of these are in current use among gin-makers, although the only essential one is the fragrant and spicy juniper berry. These berries are the female seed cones of *Juniperus communis,* a coniferous shrub or tree that is found widely across the Northern Hemisphere, though the berries used in gin usually come from Tuscany or Macedonia. Each berry contains several fused scales containing seeds that are packed with the host of different monoterpene hydrocarbon molecules (such as α-pinene, the source of the quintessential "pine forest" scent) that give juniper its amazing array of aromas. Those aromas are valued in culinary contexts that extend way beyond gin, and they range from lavender and citrus, to heather and camphor, to turpentine and resin. In the case of gin, the molecules responsible not only impart flavor but also such imponderables as mouthfeel and length. Juniper is, in other words, the backbone of any true gin.

That said, it is not unusual for a modern gin to contain two or three dozen different botanicals, in various delightful combinations. After juniper, one of the gin distiller's favorites is coriander seeds, preferably sourced in Morocco. Those seeds contain linalool, a terpene alcohol that is produced by a variety of plants and used by humans for a variety of purposes. As a flavoring for gin, linalool provides an aromatic citrus-to-lemon-grass quality, with herbal notes beneath. Citrus flavors may also be introduced into gins both from more familiar sources, such as the rinds of oranges, lemons, and grapefruits (often sourced from southern Spain), and from more exotic ones, such as the eastern Asian yuzu. Another perennial favorite among gin-makers is the dried root of angelica, a European carrot relative that, when used in gin, imparts woody, earthy tones and even green herbal notes. It is also valued by gin-makers for its ability to integrate the other botanicals used in the mix. Another addition that helps to bind other botanicals is bitter orris root from Tuscany, which is perfumed, yet earthy in character. Cassia and cinnamon are both tropical tree barks that lend exotic and highly characteristic fragrances.

The list of botanicals used in formulating gins could go on and on, but it could never be exhaustive. Every gin made has its own proprietary mix of botanicals; and because gins are increasingly produced in far-flung places, new local botanicals are being tried all the time, as in the case of the South African Inverroche Classic described at the beginning of this chapter. Not only are most of the unconventional botanicals in this gin selected from the unique and intensely local *fynbos* vegetation that surrounds the distillery; there are also probably between one and two dozen of them in this particular bottling. South Africa even has one gin-maker, Indlovu, that uses dried elephant dung to impart "the textures and flavors of the African bush" to its product. Whether this ingredient should properly be known as a "botanical" is, one supposes, as much a philosophical question as a taxonomic one.

The ancient *gebrande wyn* notwithstanding, the use of large numbers of botanicals is actually a fairly recent development. Back in the days of traditional genevers, a mere handful of different botanicals had typically been used, and this remained true pretty much throughout the nineteenth century. Nonetheless, gins continued to evolve, and the product that rang out the end of the century was very different from the one that had inaugurated it. Over the 1830s, for example, tastes in Gin Palace Britain mutated toward the "Old Tom" style (probably named for the head distiller at Boord's distillery), in which cane sugar was added as a sweetener. But as the salutary effects of column distilling began to be felt, tastes changed again, trending toward the unsweetened style epitomized today by the numerous London Dry gins on the market.

In America, meanwhile, imported genevers got a boost around the mid-nineteenth century from the arrival of fashionable mixed drinks, which in turn benefited from the greater availability of ice and the appearance of a new favorite toy: the cocktail shaker. Prohibition was also on the horizon, and when it arrived it spurred a national infatuation with gins, bathtub or otherwise. Any gin—any spirit—would do. At the same time, Americans fleeing Prohibition brought their taste for cocktails to Europe, spreading the gospel of the dry martini (see Chapter 22). Things were looking good for gin. And then came gin's nemesis, its first true grain-spirit competitor: vodka.

Until the 1950s vodka (Chapter 10), a neutral spirit made from virtually anything available, had been a strictly regional spirit limited to Russia, Poland, and adjacent areas. But after World War II, vodka began to be aggressively marketed on both sides of the Atlantic, not least by the emerging large liquor conglomerates. It was vigorously advertised as a more malleable mixer than gin, and the campaign worked: by the mid-1960s vodka was outselling its competitor, which became increasingly viewed as fusty and old-fashioned. Any gin wanting to compete in this arena had to be seen as "pure" and "clean," exactly the antithesis of everything that makes a gin interesting.

A pushback was inevitable, and initially it was clever marketing and packaging rather than a more interesting product that provided gin's riposte to vodka. In 1986 International Distillers and Vintners (IDV), owners of the venerable Gordon's, Tanqueray, and Bombay Dry Gin brands, launched Bombay Sapphire, a clear gin in a striking blue bottle. A huge marketing blitz followed, and the new product was a great success. Soon thereafter, IDV dropped the alcohol content of Gordon's London Dry to 37.5 percent ABV from the traditional 40 percent, plowing the tax savings back into marketing and thereby reviving that venerable brand as well.

As gin sales rose, others began to see possibilities. British law had favored the large distillers by setting a high minimum size for stills, and in 2008 this restriction was successfully challenged by two small distillers, Chase and Sipsmith. Soon thereafter, the makers of Adnams beer persuaded Parliament to overturn a law that had kept brewers out of distilling. As a result of such changes, by 2019 there were more than 166 distilleries in England, up from twenty-three in 2010. Most of them were distilling gin on a small scale, and thus depended for their success not on vast marketing budgets, but on the quality and uniqueness of their products. The success of this strategy is reflected not only in soaring sales figures, but more subjectively on the national pub and bar scene, where the gin and tonic is now not only the leading mixed drink, but is also available in a bewildering variety of brands of both components. The main cloud on the horizon is taxes: according to the distillers' trade organization, recent and expected rises in UK alcohol duties "will stifle the growth of the innovative, creative start-ups who have helped drive the gin renaissance."

Across the Atlantic—and indeed around the world—the story is much the same. Innovative distillers are rethinking gin everywhere. Some American distillers are even questioning the role of juniper: consider the "New Western" brands, led by the light-on-the-juniper Aviation, that some old-timers dismiss as "ver-gins."

Perhaps better deserving of this latter descriptor are the nonalcoholic "gins" such as Britain's Seedlip, which leaves out the juniper altogether and leans instead on other botanicals that supposedly evoke alcohol. This omission means that Seedlip's trio of bottlings are not strictly "nonalcoholic gins," though in intent, at least, they seem to belong in the family. In tasting them, however, we sorely missed the juniper as well as the absent alcoholic bite. They are clearly a work in progress, currently best used as mixers.

The obvious bottom line is that anything is possible nowadays in the world of gin. The beverage is rightfully taking its place as a favorite among those who are not just after an alcoholic wallop, but also want carefully orchestrated flavors—in both their cocktails, and in the underlying spirit. It may be that younger drinkers are drifting away from alcoholic drinks in general, but gin's unique gustatory qualities still herald a bright future for this lusty, gregarious spirit. The tax collector permitting, of course.

14

Rum (and Cachaça)

Susan Perkins and Miguel A. Acevedo

With the pandemic, meeting on a tropical beach wasn't possible—so we fired up our laptops for a virtual glass-clinking. We each poured a taste of Ron del Barrilito, the oldest Puerto Rico rum, its glittery golden label unchanged since the first bottle was made in 1880. The distillery was founded by Pedro Fernández, who learned how to distill Cognacs and other brandies while studying in France. Word spread of the smooth, strong spirit he produced, and soon Fernández was selling the precious liquor served from his "barrilitos," or little barrels. While rum is often consumed with mixers, as in the world-famous Rum and Coke, Ron del Barrilito is crafted to be tasted neat or on the rocks, without distraction. Poured into a snifter the color is a deep, bright amber. It smells sweet like honey or caramel in the glass, but with a bite to the nose that hints at its high proof. Despite the strength, the taste is smooth, not jagged, with a pleasant warmth that continues down the throat and complements a hot summer evening. We note hints of sweetness stemming not just from sugarcane, but also from the sherry barrels in which it spent years perfecting its flavor. It is a taste of "home" for both of us—for one, of a birthplace, and for the other, of an adopted homeland after decades of fieldwork all over the Caribbean.

Unlike the delicate wines that are painstakingly developed from the bright, luscious fruits of the grapevine, their finicky notes needing to be balanced just right and aged to perfection, rum is literally made from waste. It is produced by fermenting the product left over from refining a very useful weed—the bawdy, gangly, aggressive plant known as sugarcane. Native to Australasia, sugarcane belongs to the family Poaceae, the grasses. This family includes most of the economically important crops—several others of which are involved in the production of spirits described in this book—but, among them all, sugarcane is king. It is a colossal grass with yields of up to seventy tons of product per acre, making it one of the most efficient photosynthetic machines known. The average annual production of sugarcane dwarfs that of all other crops, with almost two billion tons grown and harvested every year around the world, primarily to produce pure sugar for sweetened foods and drinks.

The bulk of the sugarcane harvested belongs to the species *Saccharum officinarum,* native to the island of New Guinea. Austronesian-speaking peoples brought the plant northward and westward, where it often hybridized with other local varieties. As far back as 500 BCE, people in India were already enjoying the spoils of sugarcane. They developed blocks of crystallized sugarcane syrup called *khanda* in Sanskrit—a term that evolved to *qandi* in Arabic, and eventually the English word *candy.* In addition to the sugary treats themselves, the Vedas, ancient Hindu texts from about 2000 BCE, describe a cane-juice-derived spirit called *gaudi,* while the Mānasollāsa, a twelfth-century CE Sanskrit text, refers to a fermented cane drink called *asava*. Accordingly, India may well be credited with the earliest version of undistilled "rum."

As sugarcane continued its spread westward, the Greeks and Persians marveled at a plant that could "produce honey without bees." From the sixteenth century on, sugar's popularity increased throughout Europe as people grew accustomed to the sweetness it brought to bitter coffee, tea, and chocolate. Sugar remained an extremely precious commodity, however, because the towering grass could only be grown in wet, tropical climates very far from Europe.

Because of this limitation, islands to the west of mainland Europe, such as the Canaries and the Azores, became the new centers of sugarcane cultivation. This involved backbreaking work, particularly during harvesting, and the Portuguese brought enslaved Africans to the islands to do it. When Christopher Columbus, the son-in-law of a wealthy sugar merchant from Madeira, observed the islands of the Greater Antilles, he immediately recognized the alluring and profitable prospects of growing cane there. He wasted no time, bringing sugarcane plants to the Caribbean on his second voyage.

Like seemingly everything surrounding this spirit, the origin of the word *rum* is hotly debated; and frankly, any theory seems somewhat believable. Some argue that the word derives from shortening the Latin name for sugar, or from the generic name of the sugarcane plant—*Saccharum*. Others believe that it came from *rom,* a Romani word for "good" or "potent." But one of the most popular etymologies involves the words *rumbullion* and/or *rumbustion*. Their exact meanings in this context are unknown, but in England's Devonshire county they signify "a great tumult." Rum is also known as *rhum, eau-de-vie de cannes, aguardiente de caña,* or by the more colorful colonial-era nickname "Kill Devil."

The first step in sugar production is to press the cane stalks in order to extract their juice, which is then filtered and cooked. After the crystals of sucrose are obtained, what is left behind as waste is molasses (on average about one pound of molasses for every two pounds of sugar). Initially, this gooey substance was mostly an inconvenience to sugar refiners—the best they could think to do with it was to feed livestock, or to give it to enslaved workers as a caloric supplement. Then the idea struck to ferment the waste—perhaps by accident when some wayward yeasts found their way into a bucket of molasses and began to work their bubbly, alcoholic magic (see also Chapter 3). Rum, and its culture, were born.

The key ingredients of a fine rum are more than just the molasses itself, and often involve other waste or recycled products. One is dunder, the scrapings from the still left over from the previous batch, and another is water, sometimes fresh or filtered, but sometimes simply the water that was used to clean the pots following sugar boiling. (Many of the smaller volcanic Caribbean islands have scarce fresh water, so recycling was necessary.) This combination of molasses, dunder, and water, known as the wash, can be left open to be colonized by tropical yeasts in the environment that then convert the sucrose into ethanol, yielding a naturally fermented drink known as *guacapo* or *guarapo* in Puerto Rico, and as *grippo* in Barbados. The wash is then transferred to the still, where it is heated. As in any distillation process, the resulting vapor needs to be condensed to yield the precious alcohol, which is the most challenging step in the hot, water-scarce islands. In Barbados, windmills were used to power pumps that would circulate the cooling water, but in many places enslaved workers contributed most of the needed energy.

Rum production has long used both pot stills and column stills. Pot distilling, the original technique for producing rum, is a single-batch method, usually employing copper pots both for the boiling process and for collecting the distillate. As with other liquor distilling, the act of collecting that valuable product requires patience. The heads that first come off the still will be unpalatable due to their volatile alcohols and acetones, and possibly lethal because they contain methanol as well. Cranking up the heat of the boil will coax out the next distillates, the hearts of the rum that contain the desired flavors and ethanol. Eventually the distillate will begin to taper off, and the taste will decline. The tails are sometimes recycled back into the next batch to eke out a bit more spirit.

Column distilling allows for near-continuous production and will yield a significantly more consistent product, with far less energy wasted. The process was patented by Aeneas Coffey in 1830 and involved two connected columns (Chapter 2). Eventually these two-column stills gave rise to more complex stills with three or more columns, yielding more complex and individual rums. Just like in Barbados, where Caribbean rum was probably first made, rum produced nowadays on French islands such as Martinique and Guadeloupe, and known as *rhum agricole,* is made from fermented sugarcane juice as opposed to molasses. It is distilled using a "Creole" still, thought there to be the best of both worlds. It combines a pot-like boiler, the width of which corresponds to the body of the resulting spirit, with a single-column condenser.

If the rum is to be aged, this is typically done in leftover bourbon barrels imported from Kentucky or Tennessee, giving the spirit a golden color from the wood tannins. These barrels are sometimes then recharred by the rum producer to enhance the wood flavors, with the chemical compounds in the toasted wood seeping into the young rum and enhancing its flavor and color (Chapter 6). The wood itself may vary in pore size, too, which will translate into greater or lesser oxidation and the corresponding aromatic compounds. As with whiskies, sometimes different years of rum are blended to produce a consistent product for a brand. Islands with Spanish heritage sometimes employ a Solera system akin to the one used for sherries. This method stacks the barrels in rows according to age, with unaged spirit on the top row. As the rum ages, some of it is moved successively to the lower rows of barrels, and the top row receives the new batch. In this manner, the resulting rum will have a blended character.

Unlike the rules dictating the creation of bourbons or other whiskies, the regulations surrounding rum production can be as loose and carefree as a summer day on the beach—much to the chagrin of modern connoisseurs. Technically, rum is supposed to be a liquor derived from sugarcane—either molasses or cane syrup or juice—but even that basic requirement is occasionally relaxed; there is, for instance, a Kentucky-made "rum" manufactured from sorghum. The color of a rum is not even a reliable indicator of age. White rum may be aged in barrels, but then have its color filtered out, while some dark rums are fresh but tinted with colorings. Some countries, though not all, dictate that bottles be labeled with the youngest vintage in a blend; less stringent rules permit some producers to list the oldest year, or the average. Sometimes, if you look carefully, you'll see bottles bearing numbers that don't have anything to do with the rum's age! What really stokes many rum lovers' furnaces, though, is the debate over added sugar. You might think that a spirit derived from sugarcane would not need any more sweetening, but the practice of adding sugar to rum is widespread—and often hidden. Makers are not required to disclose whether there is added sugar or not, so sticklers will take their own measurements, or consult available data. Like vodka-makers, and probably in an attempt to attract a new generation of fans, rum-makers have also experimented recently with a rash of added flavorings that have expanded beyond the traditional coconut and spice to fruits like apple and watermelon.

Rum is now successfully produced in more than thirty countries around the world, but its home base and the center of its identity is firmly the Caribbean. Traveling the islands of both the Greater and Lesser Antilles, one cannot ignore the role that rum plays in both their collective and individual identities. Though rum is a ubiquitous spirit from Cuba to Trinidad, the customs and styles of drinking it vary widely across the islands. In the Dominican Republic, *mamajuana* is a popular drink. A combination of rum, red wine, roots, bark, and herbs, it is believed to have its origins in a traditional Taino medicinal tea that was combined with European alcohol in the early days of the island's colonization by the Spanish. It was enhanced with rum when production of European alcohol began. Many claim this drink has aphrodisiac properties, and variations over the years pushed that needle by adding exotic ingredients like sea turtle meat (now illegal).

Jamaicans tend to like their rum neat and strong, but you can also find fusions of rum with local fruit juices including pineapple, orange, and lime—the traditional "rum punch." For Christmas each year, Puerto Ricans make a sweet festive drink called *coquito*—a blend of coconut, rum, sugar, and spices, with each family's recipe a closely guarded secret. On the tiny island of Saba, known mostly for its spectacular scuba-diving locales, you can find a local rum-based creation known as Saba Spice. The local women combine brand-name aged rum with their own blends of spices such as nutmeg and cinnamon to yield an after-dinner liqueur that is half dessert, half drink. It is often sold in leftover "Chubby" soda bottles at local gift shops. In many a local watering hole throughout the Antilles, bartenders will lovingly tend a batch of Planter's Punch—a funky and potent blend of rum, fruits, and spices (and who knows what else)—brewing and mellowing it in an unmarked glass jar on a shelf. Order a "ti' punch" in Martinique or another French island, and you may find yourself receiving a tray with a bottle of local rum, a bowl of sugar, and a glass with a cut piece of lime and a spoon. It's up to you to adjust the proportions, depending on how you're feeling in the moment. And then there is Trinidad, where you might be dared to try a shot of strong white rum containing a bright orange and fiery-hot Scotch bonnet pepper that will leach its mouth-numbing heat into your drink.

When it comes to island rum, though, the dominant name is Bacardí, which is often synonymous with the spirit. During the colonial era, Cuba was primarily under Spanish rule. In 1762, however, the British briefly occupied the country, not even for a full year but long enough to introduce rum production, which was continued when Spain resumed its command. After a young Catalan wine merchant, Facundo Bacardí Massó, landed in the bustling city of Santiago de Cuba in 1830, he sniffed opportunity—and it smelled a lot like a new style of rum. The typical colonial rums were rather harsh and bitter to the palate (sometimes romanticized as arising from the practice of being too lazy to clean the equipment), and Bacardí endeavored to create a smoother, lighter spirit. To do this, he made three changes to the production process. He chose not to rely on wild yeasts for fermentation, but rather sought to identify and culture a particular strain that could be counted on for consistency. He opted to filter his rum through charcoal to remove impurities (though the exact process remains a closely guarded company secret). And he insisted on aging it in American white oak barrels, allowing the rum to mellow.

Following the boom in the rum business during America's Prohibition, the Bacardí company expanded to Mexico, Puerto Rico, and then, in 1944, to the mainland United States. Though the Cuban assets were seized by Fidel Castro during the Cuban Revolution, those other distilleries and resources kept the family-owned business afloat—actually more than afloat. Due in part to its numerous acquisitions of other beverage labels, Bacardí is now the largest privately owned spirit-producing company in the world, with its headquarters in Cataño, Puerto Rico. Its rum selection is now highly diversified, including white and dark varieties as well as flavored and specialty rums. (Let's not forget their now-discontinued explosive Bacardí 151—with about 70 percent ABV—which was dangerously flammable and sadly resulted in several injuries.)

Though the history of rum is not as long as those of some of the other great spirits, rum's role in the historical record is profound. Used as nutrition, medicine, and even currency, it is an essential part of the human experience. Rum is the story of colonialism, and of slavery. It evokes resourcefulness, independence, persistence, and individuality. In other words, the story of rum is the story of the Americas.

The history of rum begins with that of another spirit made from sugarcane—cachaça. Made (like rhum agricole) from the fermentation of cane juice, as opposed to molasses, cachaça was likely the first alcoholic drink to be produced in the Americas on a large scale, and to have a significant economic impact. Sugarcane plantations were established before 1526 in the Brazilian state of Pernambuco, and some historians argue that the first cachaça was produced around 1532 in São Vicente. Brazil was the most important producer of sugar and its fermented derivatives in the sixteenth century, with Dutch settlers controlling production.

During the Pernambuco insurrection of 1817, the Dutch were expelled from Brazil and settled in the Caribbean islands. The story of rum would move there too, but cachaça retained its position as the favorite spirit in Brazil—to the tune of over a billion liters consumed in the country each year. Like rum, cachaça can be white or dark, and aged or not. Most will be familiar with it as the necessary ingredient for the cocktail caipirinha, a mix of cachaça, sugar, and lime; but many locals prefer to take it straight, alternating their sips with bites of feijoada, a traditional bean stew.

Many islands claim to be the birthplace of rum, but the eastern Caribbean island of Barbados, colonized by the British in the early sixteenth century, generally gets to claim the title. Pieter Blower, a Dutch settler, brought a knowledge of distillation learned in Brazil to Barbados in 1637, and soon the island became the most important producer of rum in the Caribbean, exporting 15 percent of its production. Barbados's neighbor Martinique, a French colony, had a similar story involving the introduction in 1644 of sugar mills by the Portuguese Benjamin Da Costa, who had learned the business in Brazil. On the North American mainland, the French also tried to grow sugarcane in Louisiana, but were initially unsuccessful because sugarcane was not well adapted to its (relatively) cold winters. Manuel Solís and Antonio Méndez were eventually able to produce the state's first crop in 1794, however, using a sugarcane strain imported from Cuba. No matter who deserves the title of "first," there is complete agreement that by the end of the seventeenth century virtually every colonial outpost in the Caribbean was involved in making, selling, and distributing sugar, and therefore rum.

The harsh truth of the connection of sugar—and rum—to slavery must be stressed. Sugar, as well as rum and raw materials from the colonies like tobacco, hemp, and cotton, were brought to the ports of Europe. There, the ships were reloaded with manufactured goods such as textiles and guns, then sailed to the west coast of Africa to exchange for slaves. The ships and their horrifically overpacked human cargoes then continued their often-fatal voyages to the islands, thus completing the cycle. Many of the enslaved peoples were forced to work the sugarcane fields, replacing those who had perished under inhuman working conditions, new diseases, or outright cruelty. This three-step movement of goods and people has come to be known as the triangle trade. The number of human lives that were part of the history of rum is almost unfathomable—for example, by the time of emancipation in Jamaica, black Jamaicans outnumbered whites by twenty to one.

Although some rum-making continued in the islands, the production of rums also moved northward as ships transported the barrels of molasses to the multiple distilleries that dotted New England throughout this period. Rum production came first to Staten Island in 1664, and soon there were dozens of mainland distilleries in Rhode Island, Massachusetts, and Pennsylvania as well. These distilleries cranked out gallons of bitter rum for taverngoers and the public alike. Estimates are that the colonists—men, women, and even children—each consumed, on average, an astonishing four gallons a year.

In rum's early days, the Galenic principle of medicine was prevalent. In other words, body ailments were attributed to imbalances between hot and cold, or wet and dry, environmental conditions. Therefore, when people had chills attributed to cold and wet weather, rum was prescribed to restore body heat. It also worked the other way. When you were battling a fever (perhaps yellow fever, which was a constant plague of both the tropical islands and the mainland colonies), rum was a reputed cure. What's more, rum became intertwined with the seafaring life, both respectable and not. Because it was safer to drink than water, while offering some calories as well as analgesic and motivational qualities, rum was adopted as the official alcohol of the Royal Navy. Each sailor received a daily rum ration to the tune of an eighth of a pint, which was typically divided into two servings a day. Although the higher-ranking officers typically sipped theirs neat, the rank-and-file sailors more likely consumed theirs as grog, a combination of rum and water that was enhanced with limes if available (in 1747 citrus was discovered to be the remedy for scurvy). The practice of the Royal Naval "tot," as it was called, persisted for over three hundred years. Beginning in the mid-seventeenth century, and codified by the British Navy in 1756, these allotments of rum persisted until July 31, 1970, when their end was marked by a dramatic ceremony held in Hawaii on the HMS *Fife,* the last British ship to serve them. It seems that high technology and dangerous weapons on naval ships had made the tradition of slightly inebriated sailors too risky.

During the colonial period the Atlantic Ocean, and particularly the Caribbean Sea, were superhighways of trade ships crossing back and forth with slaves, sugar, rum, tobacco, and other goods. The temptation to steal from those ships was just too great for some, and both pirates and privateers—buccaneers paid by a government to harass and plunder foreign traders—were abundant. Because rum itself was considered a currency, pirates would have delighted at seizing a stash of the golden spirit—and they certainly would have tapped a barrel or two and enjoyed a toast to their successful pillaging of gold, silver, and other treasure. Captain Henry Morgan, John "Calico Jack" Rackham, and Edward Teach, better known as Blackbeard because of his unmistakably long, dark whiskers, terrorized the seas. Though they rarely killed the sailors of the ships that they robbed—it was far less dangerous just to intimidate them into handing over the goods without a fight—the adventures of these fearsome looters were legendary. Centuries later, this pirate connection would be used as the perfect marketing tool to associate potent rum drinks with danger and adventure, enticing customers in countless cruise ship lounges and beachside bars.

As the colonies grew, and rum continued to be intertwined with profits that funded European monarchies, the spirit also became a part of the movement for American independence. It was no doubt a potent lubricant for many a dissenting conversation in taverns, where "taxation without representation" was becoming a battle cry. The British first tried to tax and control the importation of the molasses to the colonies with the Molasses Act of 1733; but they were largely unable to enforce it. Following the Seven Years' War, the British renewed their focus on raising funds in order to support the troops needed to maintain order in their colonies. They passed the Sugar Act of 1764, which directly affected the New England port cities in which rum was produced. Outraged fledgling patriots, including Samuel Adams, vociferously objected. So although ultimately it would be the Stamp Act that tipped the balance to insurgency, war, and ultimately independence for the thirteen American colonies, the role of rum in the birth of the nation cannot be underestimated. Soon after independence was achieved, during the very first presidential election, a former general named Washington plied voters with rum as a means of encouraging their support, then celebrated his victory with a hogshead of his favorite Barbadian spirit.

Despite its role as fuel for the American Revolution, rum fell precipitously in popularity in the years following the conflict. The source of the sugar, the Caribbean islands, had not joined the mainland colonies in declaring independence, and molasses became more difficult and expensive to acquire. Rum eventually came to be seen as the spirit of the oppressor; meanwhile, as the nascent republic expanded westward, wheat, corn, and other grains became more accessible. This translated into the rising popularity of whiskey, which became the predominant American choice in spirits. But rum did not go extinct. Instead, it has enjoyed multiple rounds of renewed popularity, each intertwined with the history of the Americas.

Though fans of tropical drinks likely envision a daiquiri as something that comes out of a machine resembling a 7-Eleven Slurpy dispenser, the traditional daiquiri was handmade in Havana, Cuba. Around the turn of the twentieth century a bartender known as Constantino, at a haunt known as El Floridita, perfected the simple concoction of rum, fresh lime, sugar, and ice, which he shook vigorously and served in a stemmed glass. This tasty drink might have slipped into obscurity if not for two factors. The first was Prohibition. When the United States made the sale of alcohol illegal in 1920, thirsty Americans from the south resourcefully headed to the closest foreign port of call—Cuba. And Havana was all too happy to welcome them with open arms and cocktail shakers, sharing the delicious white rums that Bacardí and the island's other rum producers had perfected for mixers.

A famous patron further promoted the simple rum cocktail. After falling in love with the island, the climate, the fishing, the woman who would become his third wife—and rum—Ernest Hemingway spent decades living in Cuba, where he could write without the distraction of fans and cold winters. But he could not write all day and night; he ended most of his days in Havana with a few of those cocktails (and sometimes more than a few—the unofficial gossip was that on one particular evening he downed sixteen). His favorite was known as the Papa Doble—a double shot of rum, hold the sugar.

Ernest Raymond Beaumont-Gantt, originally from New Orleans, returned to the United States in 1933 after a few years sailing the Pacific. With a trove of souvenirs that could serve as décor, and Prohibition revoked, he legally changed his name to Donn Beach, opened his first Hollywood restaurant, Don the Beachcomber, and gave birth to the "tiki bar." Lounging in rattan furniture and surrounded by flowery fabrics, wooden masks, shells, and other Polynesian décor, tipplers could temporarily escape to another world. Beach's liquor of choice was rum, and he creatively crafted delicious and colorful concoctions with exotic and dangerous-sounding names like Cobra's Fang, Shark's Tooth, and Zombie. Yet it was his rival, Victor Bergeron, founder of Trader Vic's, who claimed to have created the signature drink of the tiki culture—the mai tai, which is made with two different kinds of rum (one light, one dark), lime, curaçao liqueur, almond (ideally via orgeat), and sugar. The two men fought over who had invented the recipe, with Bergeron finally winning the naming rights.

A simple Rum and Coke, a popular choice when one is first old enough to patronize a bar, also has a dose of American history associated with it. During World War II, domestic alcohol production (which was primarily gin and whiskey) ceased, so that efforts could be focused on industrial uses for ethanol. But rum continued to be produced in the Caribbean, and it served its purpose well for thirsty soldiers, as well as their supporters on the home front. When mixed with another truly American icon, Coca-Cola, the drink became practically a symbol of patriotism. The Andrews Sisters fueled its popularity when they transformed "Rum and Coca-Cola," a rather seedy and misogynistic song about prostitution, into a chart-topping harmonic ditty about the drink. Meanwhile, in Cuba, the drink became a slogan of the independence movement. The Cuba Libre—as the blend is known to this day—remains the standard in smoky jazz clubs. With a simple nod and those two words, you'll soon find yourself with bottles of local rum and cola and some glasses with ice, free to mix and enjoy as the live music treats your other senses.

What will the future bring for this robust, versatile, and adventurous spirit? In the past decade or so, rum's reputation has evolved from that of an inexpensive mixer to a respectable high-end sipping liquor, with several of the top distillers putting out expensive premium editions for connoisseurs. As rum adapts to twenty-first-century marketing opportunities, we expect more creative varieties to appear, such as the "ronmiel" from the Canary Islands that incorporates honey into its final mix. But whatever the innovations to come, a simple neat glass of dark rum will likely remain popular as an elegant classic, its original recipe practically unchanged for four hundred years.

Crossing Borders

Other Spirits, Mixed Drinks, and a Look Ahead

15

Eaux-de-Vie

Pascaline Lepeltier and Ian Tattersall

The unassuming small bottle, tied at the top with a cloth cap and a red ribbon, bore a simple, hand-lettered label with the name Zibärtle Wildpflaume. After learning of our interest in eaux-de-vie, the German friend who gave it to us had spent days hunting it down. As he explained, the spirit within had been distilled in tiny quantities in the Black Forest region, using a rare indigenous wild plum that has a large stone and only a thin covering of flesh, a distiller's nightmare. "This one is really special," he said. The label also bore the statement "Eight Years," and we wondered what that meant, since the liquid was perfectly clear and colorless. It turned out that at least some of those years of maturation had been spent in ash-wood barrels that impart virtually no color or flavor themselves, but allow the fruit flavors to integrate. Back home we opened the bottle and poured eagerly, not sure what to expect. When we swirled our glasses, a gentle fragrance rose into the air, followed by mouth-filling flavors that had something in common with plums, but were also curiously herbal. The warm, soft finish lasted for minutes. However little of this gem its maker produced, the task was undertaken with extraordinary artistry and skill.

Eau-de-vie is a tricky term to pin down, because it's French for "water of life," a description applied to distilled alcohols of many kinds and in many languages. As used by anglophones today, it is a contraction of the French *eau-de-vie de fruits,* a term that in theory indicates any alcohol distilled from fruit, but that in practice is used more specifically. Favorite fruits for distilling vary regionally: in France pears probably top the bill, in Germany cherries are often the choice, and in central Europe damson plums are widely favored. But in this slippery category, almost anything goes. In the Caribbean (and even at one distillery in Alsace, using Tahitian ingredients), guavas, mangoes, and pineapples are also distilled directly.

While grapes are undeniably fruits, distilled grape products are usually thought of as brandies rather than as eaux-de-vie (for more on brandies, see Chapter 9). As, indeed, are some more conventional fruit distillates: because it is typically aged in oak barrels, for example, the French Calvados is usually considered a brandy, even though it is made from apples. In this chapter we will confine ourselves to eaux-de-vie in the narrowest sense of the term: clear, unsweetened spirits distilled directly from (rather than flavored by) non-grape, non-grain sources, which are usually fruit but occasionally nuts and even root vegetables. We will also confine ourselves mainly to the French and Germanic heartlands of eaux-de-vie. (For various fruit spirits from elsewhere, see Chapter 21.)

Because eaux-de-vie are so widely made, the category as a whole probably deserves to be considered alongside the "big six" of mega-sellers described earlier. But we have chosen to discuss them among the more specialized distillates because they are rarely made on a large scale. The overwhelming majority of eaux-de-vie that one encounters—other than in airport duty-free stores—are made by very small producers who proudly present their wares in bottles bearing ostensibly handwritten labels.

The small-scale, artisanal aspect of most eaux-de-vie is particularly appropriate to the character of these spirits because, if any category of distilled beverages can be said to possess terroir, it is eaux-de-vie. Whereas brandies and whiskies depend on barrel aging to augment and deepen their flavors, the makers of eaux-de-vie strive instead simply to capture the essence of the ingredients from which they distill; and the qualities of those ingredients, mainly fruits, are intensely local. Each region, each orchard, and each local variety of fruit has its own particular history and characteristics; and the fidelity of an eau-de-vie to its unique origin (geographical as well as botanical) is virtually definitive of its quality. Even here, though, we run into partial exceptions. In Alsace, the heartland of eaux-de-vie and the home of the biggest producers, the production process is strictly regulated but there is no legal requirement for the fruit to be locally grown. Still, this doesn't mean that the fruit is not of high quality, or even that it is lacking in terroir: traditionally, the mirabelles (a variety of yellow plum) used in Alsace come from trusted producers in next-door but slightly cooler Lorraine; many of the pears are sourced from equally trusted growers in the Isère Valley, near Lyon; and apricots come from the Mediterranean Languedoc-Roussillon region—even though, for reasons of cost, fruits are increasingly sourced from other European countries, especially Poland and Hungary.

The distillation of fruit goes back to the very origins of distilled beverages in Europe, and it has a separate history in every area of the subcontinent in which it is practiced. Derivative colonial traditions can, of course, largely be traced back to their respective home countries. Everywhere, perfectly ripe fruit (optimum ripeness is critical) is crushed, pressed, and fermented. This doesn't necessarily mean that the fruits are picked fully ripe; pears, for example, are routinely picked early and then allowed to rest in cellars until they have reached their ideal maturity, while apricots have to be picked just before they are ready to fall off the tree. For the distillation itself, copper alembic stills are typically used, with two passes of the distillate (except when the fruit has previously been macerated in alcohol). The primary distillation raises the ABV of the fermented fruit wash to somewhere around 30 percent, and the second takes it up to about 65 or 70 percent, from which it is usually diluted with spring water to between 40 and 45 percent ABV. After distillation, the spirit is allowed to rest in inert stainless steel, pottery, or glass (or very rarely in those ash-wood barrels), when it will often naturally lose some of the excess alcohol.

It is in Alsace, in eastern France, that the production of eaux-de-vie is most intensive, and it is certainly where the eau-de-vie in your local liquor store is most likely to have been made. Under the influence of neighboring Germany, the region quickly developed a tradition of distilling local (and eventually farther-flung) fruits, in addition to the grape pomace left over from wine production—and indeed, almost anything else you might imagine distilling, from rowan berries to unusual herbs and even pine buds. Distilling became a domestic matter as well as an industry, and at one point the average Alsatian family was consuming more than a hundred gallons of its own product every year.

Perhaps it was this eye-popping level of consumption that led the authorities to change the rules in 1953, subjecting distillation to a heavy tax. Families already involved in distilling were exempted from those taxes for the lifetime of the family member in whose name the newly required license was issued—usually the youngest, and preferably a newborn if one was available. Ian's sister vividly remembers spending a year in Strasbourg during the 1970s, during which she befriended one of those families. Their permit allowed them access to a vital linking component of the still for a mere two weeks of the year. During this short period, all available hands distilled night and day before handing the link back to the authorities—and then they collapsed, exhausted. With this kind of commitment to distilling, the Alsatians became legendary for the quality of their product.

Quoting the sommelier at one of Strasbourg's leading restaurants, the *New York Times* writer and epicure R. W. Apple Jr. once memorably described an Alsatian eau-de-vie as showing both "finesse and power," two opposites "not often found in the same glass." This unusual combination results from the ability of a good distiller to capture the essence of a fruit with both high definition and weight in the mouth; and it is typical of the best Alsatian eaux-de-vie. If you are looking for purity in your spirit, rather than complexity, this is one of the best styles to sample.

But not all Alsatian eaux-de-vie follow the same production path. Although most are double-distilled in those copper alembics, some are distilled only once. The largest modern eau-de-vie producer, G.E. Massenez, established market dominance back in 1913 by impressing the Queen of Sweden with its wild raspberry eau-de-vie. Raspberries had always been difficult to distill because of their low sugar content, but Eugène Massenez figured out that macerating them in a wine-derived neutral spirit early in the process would compensate for this deficiency. What is more, raising the initial alcohol in this way would make a single distillation adequate to purpose, while preserving more of the flavor-producing congeners. The company never looked back. Just for the record, according to Apple it takes G.E. Massenez eighteen pounds of wild raspberries (nowadays mainly from Romania) to make a single 700 milliliter bottle of its framboise. In contrast, a mere ten pounds of sugary mirabelles will do the trick, although if you are making a poire it will take you thirty pounds of pears—plus one, Apple points out, if you want to make a bottle with a whole pear inside. This fat "prisonnier" ends up in its cage because the bottle was hung on the tree while the fruit was still a slender blossom that would happily flourish and develop into a fruit inside its private greenhouse. Word is, though, that while the result looks impressive, the pear has little if any effect on the spirit that will eventually join it in the bottle. The kind of pear is, by the way, critical; the preferred variety in Alsace is the Williams, which is actually the familiar Bartlett, originally bred in England and now a favorite of U.S. distillers too.

With its strong reverence for tradition, Alsace kept the production of eaux-de-vie alive in the postwar years when French consumers were turning away from this brilliant family of spirits in favor of imbibing Scotch (indeed, France became the world's largest market for Scotch during this period). Thanks to a cluster of distilleries located in the "spiritual heart" of the region—Ribeauvillé in the Haut-Rhin, and Villé in the Bas-Rhin, both with terroirs of west-facing slopes of schists and clay, perfectly suited for orchards—local camaraderie and pride preserved the craft and the excellence of the products while at the same time opening the way for innovation. Today, with their remarkably extensive ranges, Metté, Windholtz, and Holl have repositioned this category as both trendy and original, and as a base for cocktails as well as for sipping. Metté makes an impressive asparagus distillate, as well as a top-selling ginger variety, a garlic version, and eighty-five other eaux-de-vie. To reach the requisite quality, these firms prefer to use small traditional French *alambics à repasse* rather than the more modern German column stills, even though the pot stills are much more labor intensive. They also don't compromise with post-distillation aging: twelve months of rest time is a minimum for most of them, but this period may last as long as twenty-five or thirty years for some fruits like the wild blackthorn.

We can't leave Alsace without paying tribute to the oldest official appellation in France for a pitted-fruit spirit (the appellation d'origine contrôlée, or AOC, regulates production of a food or drink considered part of the French culture). The Kirsch de Fougerolles received its own AOC in 2010 and has since become available again in the traditional hand-blown, dark-glass bottles known as *bô*. From neighboring Lorraine, the famous Mirabelle de Lorraine received AOC recognition only in 2015; but in 1953 it was the very first such spirit to be recognized with a regional appellation (AOR). According to legend, the mirabelle tree was brought to Lorraine from Provence in the fifteenth century by René of Anjou. It has certainly thrived in the area since that time—as we have seen, even the fussy Alsatians prefer their mirabelles to come from Lorraine. Seventeen producers in the region now distill the fruit of the two mirabelle species authorized there for spirits production: the larger and more colorful mirabelle of Nancy, and the smaller, sweeter mirabelle of Metz that grows only on limestone-based soils and must by law be made in pot stills.

Eaux-de-vie de fruits are, however, far from a monopoly of eastern France. They are distilled all over the country, enabling one to appreciate just how clearly an eau-de-vie de fruits can show off its terroir. Significantly, a lot of recent interest in this category has come from winemakers, or from distillers who remain in close touch with the wine world. Burgundy, for example, is a region more famous for its wines, and for the fruit liqueurs made there (usually on a base of high-proof spirit, the most famous among them being the notorious *crème de cassis* that Canon Kir, mayor of Dijon from 1945 to his death in 1968, employed to improve his lean, bright Aligoté wine), than it is for its fruit distillates. But the famous Meursault wine producer Jean-Marc Roulot is continuing a family tradition of distillation that started as long ago as 1866, and his raspberry and pear spirits (from the Hautes-Côtes de Beaune) have entranced sommeliers worldwide. In the southwest town of Gaillac, Laurent Cazottes is renowned for recognizing the quality of the local fruits and the importance of how they are grown; in the process he, too, has seized the attention of the wine world. He sources only organic and biodynamic heirloom varieties that are mostly grown in his own orchards or those of friends, enhancing those sensory qualities that create the feeling of terroir: after harvest, he uses basically the same techniques that are used to produce "straw" or "raisin" wines, allowing the fruits to desiccate on clay beds to concentrate the aromas and slightly oxidize. Especially in the case of pears and greengages, he usually removes the pips, stems, and skins by hand, keeping only the flesh. Positioned at the opposite end of the spectrum from their refined Alsatian counterparts, his powerful, textural eaux-de-vie have gained recognition around the world, and you can spot Cazottes Distillerie bottles—including its astonishing "72-tomato" liquor—in many of the trendiest restaurants and bars around the world. But these are just a couple of examples: today, every fruit-growing region of France is looking back to and relearning its own local heritage (with the remarkable Distillerie du Petit Grain in the Languedoc, for example, producing what is probably the country's best apricot eau-de-vie). In a country renowned for its timeless traditions, this amazingly forward-looking, creative spirit is an inspiration.

We will not dwell here on neighboring Germany because the distilling tradition there is considered in Chapter 16. Suffice it to note that in that country a vast amount of fruit is distilled, destined for a wide variety of spirits and liqueurs that include some exquisite eaux-de-vie. German fruit spirits are sometimes referred to as "Obstlers" and are distilled mainly in the south, including by such heavyweight Black Forest producers as Schladerer and Fidelitas. Slightly farther east, Austria is home to some of the most innovative eaux-de-vie creators anywhere. In many areas there is a lively tradition of home distilling, with farmers typically producing twenty or thirty liters a year of *Hausbrand* for family consumption. Out of this tradition came Hans Reisetbauer.

Reisetbauer, who came into prominence in the early 1990s, believes that the original fruit accounts for 50 percent of the quality of the final product, and accordingly insists on growing his own when he can. Most of his production is of an exquisite Poire Williams; but he grows and distills eight other kinds of fruit. He has even recently added an orange spirit to his range—although, because he ages it for four years in oak casks, it doesn't fall within the definition of eaux-de-vie that we are using here. But other exotic Reisetbauer products do: the distiller is famous for his carrot eau-de-vie, which you will have to taste for yourself to appreciate how it captures the essence of this humble vegetable. The carrots are sourced from a local farmer, and reportedly ninety pounds of them are required to produce a single bottle of spirit. Other unusual entrants in Reisetbauer's lineup of eaux-de-vie include ginger and hazelnut, both of which are quite commonly macerated in spirits, but are rarely directly distilled.

Reisetbauer is far from the only game in town. Also very highly regarded is Destillerie Purkhart's "Blume Marillen," an apricot spirit from Austria's Wachau region that famously lingers on the palate. Also not to be forgotten is the outlandishly priced range of eaux-de-vie from the Tyrolean producer Rochelt, which rewards your investment not only with greenish crystal glass bottles, but also with remarkable finesse, especially given their whopping 50 percent ABV. The lineup from Rochelt includes not only some masterful traditional eaux-de-vie such as elderberry and quince, but some unusual combinations as well, such as the Hollermandl, which combines elderberry and pear. Other Alpine producers include Morand, located over the border in Switzerland, whose "Williamine" poire, in particular, has long been prized among aficionados. Also in Switzerland, the long-established Etter Zuger is admired for its kirsch, made from mountain-grown black cherries, while the almost equally venerable Fassbind Pflümli makes a widely appreciated spirit from *Löhrpflümli,* Swiss sweet plums.

Perhaps inevitably, the making of eaux-de-vie has made its way not just across Europe but also to the United States. All over the country, American craft distillers are making clear fruit spirits of often remarkably high quality, although the market is still dominated by two large brands. Using small copper pot stills, California's St. George Spirits makes highly regarded pear and apple spirits that they advertise as "brandies," but that fit our definition of eaux-de-vie (though its barrel-aged apple product doesn't). And in Oregon, Clear Creek Distillery makes a broad range of delicious eaux-de-vie that conventionally includes pear, cherry, blue plum, and raspberry (including a pear-in-the-bottle), as well as Douglas fir, an aromatic product made by gathering fir buds in the Spring, infusing them in spirit, and redistilling them with an infusion of new buds (which gives the result a pale greenish hue). That process may or may not disqualify this Clear Creek product from being an eau-de-vie as most narrowly defined, but it closely recalls the *eau-de-vie de bourgeons de sapin* (pine-bud spirit) that is made by a handful of producers in Alsace.

There is a lot of disagreement over exactly when eaux-de-vie ought to be drunk, though the answer should surely be "when you feel like it," which is certainly the case in most places where they are produced. If you visit a farmer who distills, for example, he or she will almost certainly invite you to sample a welcoming glass, no matter the time of day. In more formal contexts, eaux-de-vie are occasionally offered as apéritifs; traditionally, they were enjoyed in the middle of long, formal French meals to revitalize the appetite (the tradition is known as *coup du milieu* or *trou normand,* with regional variations). Today they are more commonly consumed as digestifs, after the meal—though many will argue that if a fruit-based dessert is served, the appropriately matched fruit spirit should be served alongside it.

The optimum service temperature is also debated; ideally, it will depend on the individual spirit. Some sophisticates like to keep the bottles chilled in the freezer. Poured into a chimney glass such as the Riedel Veritas Spirits model—itself stored at room temperature—the spirit will be at the perfect point to tame the alcohol while revealing the aromatics. Normally eaux-de-vie are best consumed just below room temperature, which is certainly what you'll want if your drinking style is to toss them back; but some sippers would rather receive their spirit a bit colder, and then appreciate how it evolves as it warms up in the glass. There are even those who prefer to experience the process in reverse, by slipping in an ice cube after the first sip. In that case, of course, you will need a reasonably capacious glass—something that will allow the aromas to rise from the spirit irrespective of any tinkling ice. In our view, the shot-sized glasses in which eaux-de-vie are often served rob the drinker of the most complete experience these venerable spirits can provide. Still, however and whenever you prefer to enjoy your eau-de-vie, there is no better way to end a good meal.

16

Schnapps (and Korn)

Bernd Schierwater

The dark-green rectangular bottle bore the portrait of Otto von Bismarck, Germany's famed nineteenth-century "Iron Chancellor" who, though no social democrat, implemented the country's first social laws. Bismarck's family had run a major distillery since 1799, so the name on the bottle's label seemed to promise the drinker a well-made Korn as well as a historic, princely, and majestic experience. In the end I wasn't disappointed: I could almost taste in this clear spirit the principle that good things will survive and can offer a world of pleasure even after two hundred years of following the same recipe, the same tradition. This was a classic clear-water Korn with a smooth and mild attack, a delicate mingling of flavors, and a beautiful clarity and purity. As I picked up my Schlückglas and admired its contents, there really was only one thing to say: *Prost!*

My ancestors were experts at distilling clear schnapps, hence our family name Schierwater, or "clear (alcoholic) water." In its strictest sense, the drink they perfected, clear schnapps, is the same as Korn, a distillate from grain, while other kinds of schnapps are made from other plants, especially various fruits. Such fruit distillates are known in Germany not only as Schnäpse, but also as Obstlers, which although very different from Korns, are similarly clear in appearance. Any reader who would like to learn more about Obstlers should consider visiting the annual Stanz brennt festival in the formerly German town of Stanz, in Austria's Southern Tirol. Despite the town's small size—Stanz has just 650 residents—you can try Obstlers from more than fifty different distilleries.

Most people in Germany use the term *Schnaps* not only for clear Korn, as I do here, but also for anything from Obstlers to vodka, Cognac, and whisky. This imprecision reflects the decline in popularity of slow, simple, high-quality, clear Korn in a modern society that is driven by speed, diversity, and phony ephemerality. These days, consumer products seem to sell better if they are colorful, fancy, and artificial, whereas the traditional Korn is anything but. It is clear and straightforward and has not changed much in five hundred years. A Korn has a minimum of 32 percent alcohol (64 proof), is only made from defined grains, and may not contain any additives. The only grains allowed are rye, wheat, buckwheat, barley, and oats. In practice, rye and/or wheat are most often used nowadays, while barley is used for malt production and buckwheat and oats are only rarely distilled. By law Korn, and its classier sister Doppelkorn, may be produced only in Germany, Austria, and a small German-speaking corner of Belgium.

Although the spirit was already being distilled much earlier, the first historic mention of Korn is found in a tax document issued in 1507 by the city of

Nordhausen, in the Free State of Thuringia. Within forty years, however, in 1545, the first Korn prohibition had been imposed. This was apparently at the behest of the beer brewers, who found that the competition for grain was making barley too expensive for them. Still, Korn was much too important to society to be suppressed, and the prohibition lasted only until 1574, when Korn was once again free to become important economically. It hit a short recession during the Thirty Years' War (1618–1648), but subsequently went on to assume increasing social and economic importance over the next three hundred years, with some stumbling blocks along the way. In the middle of the eighteenth century another kind of clear schnapps (Klaren) began to be produced from cheaper potatoes, resulting in 1789 in the proclamation of the first *Reinheitsgebot* (German "purity law") for Korn, in which the city fathers of Nordhausen declared that it should contain at least two-thirds rye and not more than one-third barley. One of the most famous Korn distillers in that period was the father of the Imperial Chancellor Otto von Bismarck, whose family operated a distillery. Its label "Fürst Bismarck" is still one of my favorites, even though this Korn now comes in a green bottle instead of a clear one. Because of continuing competition from the cheaper potato distillates, in 1909 a broader Reinheitsgebot for Korn was declared throughout Germany.

During World War I valuable metals such as copper, brass, and bronze were confiscated for the war effort, so that all distilleries closed except for the few needed to distill Korn for the military. In 1924 Korn came back briefly again, before being prohibited by the Nazis in 1936. Only in 1954 was this prohibition ended by the Federal Republic of Germany, leading to a new renaissance. Production quickly boomed, and Korn regained its popularity. In the 1960s some producers added apple juice to their Korn, to make an "Apfelkorn." This became an oddly popular drink during the Flower Power era, although most drinkers remained true to the traditional Korn.

At special events such as wedding parties and anniversaries, a special *Doppelkorn* is often served. Whereas regular Korn is distilled once, Doppelkorn is distilled up to seven times and has a higher alcohol content of 38 percent ABV (76 proof), compared to regular Korn's 32 percent (64 proof). As a result, Doppelkorn is very clean and classy. When my father became a grandfather for the first time, he opened a bottle of it. The nobler Doppelkorns are matured in wood at around 85 percent ABV (170 proof), before being diluted with water to the final drinking concentration of (usually) 38 percent. The quality of the water used is important, and mineral-free spring water is normally employed. Where melted glacier water is used, the Korn is called Eiskorn (ice Korn). Together, the quality of the water, as well as the number of filtrations and distillations, determine much of the taste of a Doppelkorn. As with regular Korn, no additives of any kind are allowed. Sadly, though, the concepts of high quality and high purity seem to be losing influence nowadays, and both Korn and Doppelkorn have lost market share in Germany. Fancy drinks with fancy names, unknown ingredients, and artificial colors sell better, although recently the list of cocktails that use Korn as an ingredient has been growing.

Is it true that a schnapps helps digestion after lunch? No. A glass of schnapps helps digestion after breakfast and dinner as well. The spirit activates the production of gastric acids and stimulates enzymes for protein digestion. Until the 1970s a bottle of Korn, Doppelkorn, or Steinhäger would invariably be found in any good restaurant or home bar in Germany. Steinhäger is a Doppelkorn that is produced in the city of Steinhagen and sold in stone bottles (*Stein* means stone). Where I grew up, we always had Steinhäger at anniversary parties. It was also inconceivable to play cards without glasses full of Korn, Doppelkorn, or Steinhäger—or, indeed, for a laborer to consume a ground meat roll on the job site without a schnapps. A Korn after breakfast helps to get the body back to work; a Korn after dinner helps you sleep better. How things change: nowadays, Brazilians drink more than twice as many bottles of Steinhäger as Germans do.

Some of the traditions have continued, however. On Ascension Day (also *Vatertag,* Father's Day), fathers and prospective fathers derive from Korn much of the energy needed for pulling a *Bollerwagen* (handcart) filled with beer and schnapps for hours and miles through the parks and woods. This peculiar way of celebrating Father's Day goes back to the Middle Ages, when workers were allowed to celebrate the arrival of summer by joining in the religious processions. After Martin Luther initiated the Reformation in 1517, the ritual changed and praying was replaced by beer, schnapps, and the company of unmarried women (the wives had to stay home to take care of the children). Nowadays, in some locations gentlemen's Vatertag parties are arranged commercially with nicely decorated carriages, and sometimes even without Korn. In 2008 the German minister for families, Ursula von der Leyen, called for the impossible: alcohol abstinence on Father's Day. She apparently did not realize that a Father's Day without Korn is like a Thanksgiving without turkey.

There are currently some eight hundred Korn distilleries in Germany, producing a couple of thousand different varieties of the spirit. Here are some of my favorite Doppelkorns:

- Echter Nordhäuser. This traditional distillery in Nordhausen has been making Korn since the very earliest days and produces what is probably the best-known Korn of them all. The distillery uses only rye for grain, and, if necessary, German oak barrels for aging.
- Berentzen. The biggest Korn distillery is Berentzen in Haselünne. It uses only wheat, and the aging in oak barrels lasts at least twelve months. Alcohol content is a uniform 38.5 percent ABV, and in addition to the standard 0.7 liter bottle you can also buy its Doppelkorn in attractive three-liter bottles.
- Fürst Bismarck. The successors of the Korn aficionado and Imperial Chancellor Otto von Bismarck opt for mild wheat, combined with strong rye.
- Hof Ehringhausen. Wheat alone, oak barrels, and an intense 42 percent alcohol content are the characteristics here. The Doppelkorn is also sold in small 0.2-liter and 0.5-liter flasks.
- Hardenberg. The Hardenberg distillery in Nörten-Hardenberg is famous for its *Hochzeitsweizen* (wedding wheat), a well-aged Doppelkorn that was originally made as a wedding gift for the daughter of the aristocratic Carl-Hans Graf von Hardenberg.
- Ostholsteiner. This very clear Doppelkorn has been distilled since 1824 in Lütjenburg. Made from wheat, it is filtered nine times and diluted with special water derived from the frontal moraine of a glacier.
- Berliner Brandstifter. Since 2009, each bottle of wheat Doppelkorn made here has been hand-bottled and hand-signed.
- Schwarze & Schlichte. This distillery began its commitment to Korn more than 350 years ago and matures its flagship Friedrichs Fasskorn in oak barrels previously used for aging Cognac, whisky, or sherry. Two years in barrel adds significant flavor and color to the Doppelkorn.
- Klosterbrennerei Wöltingerode. In 1682 the Klostenbrenner monastery was forced to distill and sell Korn to pay for rebuilding after a fire. Its special

Edelkorn, which uses only hand-ladled spring water from the monastery's well, is dedicated to the monastery's famous provost Johannes Wapensticker.

- Kornbrennerei Warnecke. This distillery in Bredenbeck offers tours that show visitors how the Korn (grain) becomes Korn (schnapps).

The production of Doppelkorn follows a broadly similar path to that of whisky, but it is a unique spirit with its own, more robust personality. Similar products are made outside of Germany and Austria, but are not allowed to use the protected Doppelkorn name. Every ingredient, every manufacturing step, and every hour spent aging in wood (or sometimes pottery) vessels is important to the spirit's taste, and so is the art of drinking it. Some people like to drink any Korn at room temperature; others like it ice cold from the freezer. You can enjoy less or more flavor and aroma by shooting your schnapps or sipping it. But whatever your preference, you should always drink it from small, two- or four-centiliter Schluckgläser (0.7- or 1.4-ounce shot glasses). Since Korn has more flavor than vodka, it has replaced vodka in many German cocktails. In fact, both simple Korn and the nobler Doppelkorn go well with a wide spectrum of soft drinks, but since my name is Schierwater and I drink my Korn straight up, I skip the many possibilities this precious spirit offers for upgrading other beverages. By the way, you can keep your Doppelkorn for several decades, possibly much longer. But even if you stock up, I guarantee you will never discover that for yourself.

17

Baijiu

Mark Norell

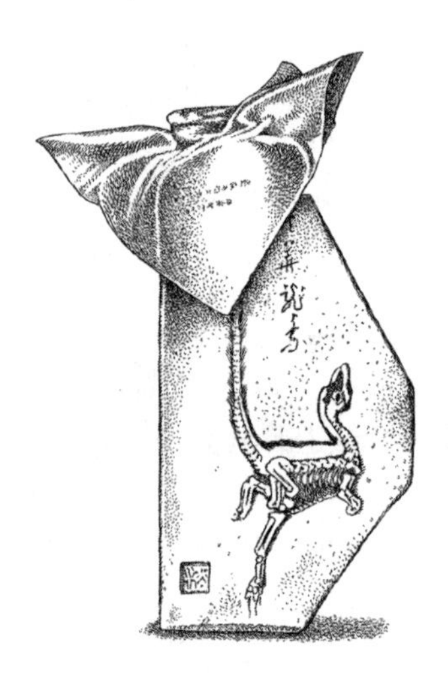

Baijiu is so ubiquitous in China that commemorative bottles are inevitably produced. And there is a lively aftermarket for them, not only among genuine collectors, but also among counterfeiters who might fill a particularly rare vessel with some low-grade spirit, secure in the knowledge that the bottle is likely destined not for consumption but for a place of honor in a glass case or on a shelf. The internet does a lively business in such items, with some empties going for several hundred dollars. This commemorative bottle is from the northeastern Chinese province of Liaoning, home to many a baijiu distillery.

It is also the area where the first feathered dinosaur fossils were found beginning in the 1990s. It was presented to me around 1999 in Beipiao, a town in central Liaoning that hosted an early conference on the region's extraordinarily rich and scientifically important fossils. The dinosaur portrayed here is the first feathered dinosaur to be discovered: a small form, about two feet long, named *Sinosauropteryx*. Its name means "Chinese reptile feather," but the inscription on the bottle says "Chinese dinosaur bird alcohol." The baijiu within was a rice variety, but my recollection of its taste is delightfully hazy, since it was consumed over a very long and boozy celebratory evening.

Ever heard of baijiu? For me, the name conjures up some of my fondest memories of China, and some of the worst. Hands down, it is responsible for the worst hangovers I have ever had. Baijiu hangovers are so bad they have been known to break blood vessels in the eyes and catalyze a week's belching. But for all this, baijiu is loved in China. It is the national spirit of the country, available to both the rich and the poor. Which means that, due to the sheer size of China's population, its ancient tradition of festive and social drinking, and very few Chinese cultural or religious mores regarding ethanol consumption, it is the most popular spirit in the world: more baijiu is consumed worldwide than whisky, gin, tequila, and vodka combined. It can also be the most expensive: in 2011 a single bottle sold at auction in China for the equivalent of $1.5 million.

The Western lexicon is insufficient for describing the many flavors of baijiu. The first words that come to mind are "strong," "warm," and "funky"; indeed, many of the metaphors used by Westerners to characterize it are at least mildly pejorative. The term *baijiu* itself roughly translates to "white liquor," which is pretty general—but then again, baijiu is not just one thing. Like whisky, it is a collective term for a lot of different regional styles of distilled spirit. The varieties are known as "fragrances" or "aromas," terms that broadly correspond to "styles" or "flavors" in the Western lexicon. The lightest and most appealing to the Western palate is *Mixiang* (rice aroma), produced primarily in southeastern China. Departing from most other baijius, it is made primarily from rice, and has a delicate feel on the palate. When you sniff it, you get a whiff of rice on the nose. Another light variety is *Qingxiang* (light aroma). Very popular in the north of China, it is distilled from sorghum fermented with a barley- and pea-based starter (more on this later). And the most popular working man's baijiu is the light-aroma *Erguotou,* sold in small green bottles that are ubiquitous all over the country. Scores of pocket-size empties fill the airport trash bins just before the security check, discarded by nervous travelers who needed a little liquid courage before flying.

Baijius from the south are the hardcore traditional ones. The fragrances here are *Nongxiang* (strong aroma), and *Jiangxing* (sauce aroma). Sauce aroma baijiu is very hard to make and goes through several distillation and fermentation stages to produce a drink that is not for the faint of heart. I remember reading a Chinese airline magazine that described them as "not recommended for greenhorns" and "only for genuine drinkers." These baijius are largely produced in the provinces of Szechuan and Guizhou, regions of China known for strong flavors and incendiary hot food; in these regions, strong and spicy spirits are considered an important element of any spicy meal. Indeed, to say that the sauce and strong baijiu aromas are pungent is a huge understatement. An American colleague described a cup of the strong-aroma baijiu as tasting like "vomit mixed with gasoline, drunk strained through a footballer's used sock." Nonetheless, in China the strong-aroma and sauce-aroma varieties include some top-shelf bottles. Brands like Wuliangye (strong) and Moutai (sauce) are the Rolls-Royces of Chinese spirits. Contemporary production bottles of Moutai are priced as high as several thousand dollars, while collector bottles have sold for more than a million.

Like many other distilled spirits, baijiu is produced by fermenting cereals. In the case of most grain-based spirits, converting grain to drinkable alcohol is a two-step process, in which mixtures of cereal and water are combined and heated to achieve saccharification, that is, conversion of starch into the sugars that are then fermented. By contrast, baijiu production is accomplished in a single step, whereby a block of grain-based starting culture of microorganisms (called *qu,* or "mold cake") is mixed with steamed cereal. The qu includes all the enzymes for saccharification, as well as the yeast and other microorganisms necessary for fermentation. The mix of cereal and qu is left to ferment for up to three months, depending on the style desired and the seasonal temperature of the factory.

Sorghum is the most popular cereal for making baijiu, but other grains such as wheat are sometimes used. In the cool, temperate northern regions of China, light-aroma baijius are made from sorghum, and fermentation usually takes place in large ceramic vats that are buried underground to provide a constant low temperature. In the south, rice-, strong-, and sauce-aroma baijius are distilled almost entirely from sorghum, and are fermented in clay- or stone-bottomed fermentation pits. Those pits are lined with stone, clay, or occasionally brick, according to region, and impart distinctive terroirs.

The fermented product is then distilled, traditionally in a contraption that looks like an outsize vegetable steamer (now usually made of stainless steel), to a proof of up to 140. Before bottling, the distillate is diluted to a market proof of around 105, although the strongest may barely be diluted at all. As is true for most spirits, the water softens the taste and unleashes many of the volatiles. Almost all baijiu is blended by master blenders, in a fashion reminiscent of blended Scotches, and each blender (and label) has a distinctive style and taste. The distilled baijiu is usually then aged in large ceramic, limestone, or stainless-steel vessels, for a period of no less than a year. Sauce-aroma baijiu requires more than three years of aging.

So much in history has obscure origins, and so it is with most spirits including baijiu. One story from antiquity holds that, during the Zhou Dynasty (1046–256 BCE), a man named Du Kang stored sorghum seeds in the cavity of a hollow tree, where they got wet. When Du returned, he smelled a remarkable scent that inspired him to make more of the delicious potion. Although distillation was yet to come, this beverage is said to have set the stage for sorghum-based alcoholic drinks in ancient China.

There is some archaeological evidence that Chinese people were distilling spirits as early as the Han Dynasty (202 BCE–220 CE): scenes that might be of distilling are clearly impressed into bricks of the period. Additionally, several bronze "stills" from this time are preserved in the Shanghai Museum, though their exact function is a matter of debate. The first modern-style baijiu is sometimes considered to be a Tang Dynasty (618–906 CE) innovation: the Tang poets Bai Juyi and Yong Tao gleefully describe drinking something that might be of the same general kind. But it was only in the Yuan (1279–1368 CE) and Ming (1368–1644 CE) dynasties that the manufacture of baijiu took on its modern form, as new and more efficient distilling technology came in from the Middle East, spurring the adaptation of traditional Chinese stills to mass-production methods. During the Ming Dynasty, baijiu was also incorporated into the *Bencao Gangmu* (the Chinese *Materia Medica* of the time).

Considering how much baijiu is produced and ingested, it is strange that it has yet to develop an international following, especially in an era when everything from the peatiest Scotch to the bitterest amaro has found its way into the world's tony cocktail bars. The makers of baijiu still have not figured out how to enter the Western market in any significant way, although the prevalence of Chinese tourists and expats is making it more visible, at least locally. A few years ago one of the preflight watering holes at the Newark airport offered a small selection of baijiu brands; and growing Chinese influence, together with a fascination with all things Chinese, has led to the opening of baijiu-based bars in New York City, San Francisco, and Los Angeles. One New York City establishment, called Lumos, featured lots of different high- and low-end baijius, as well as a broad selection of baijiu-inspired cocktails. But even though one of its publicists advertised the aromas of baijiu as "the smell of sex," apparently it was not the kind of sex that New Yorkers prefer (which covers a pretty broad spectrum of activities), since the bar is now a memory. Baijius can, however, be found regularly in American urban Chinatowns.

There are literally thousands of different brands of baijiu made in China. Some are large operations and distribute their products internationally, like Wuliangye or Moutai (which is based in the town of Maotai, the moniker by which one kind of baijiu is best known); others are from small regional distilleries or from mom-and-pop-style village moonshining operations. Here are just a few at random. A 375 ml bottle of classy Kweichow Moutai will set you back around $350, while a top-end bottle from this producer can cost a few thousand. Many of its best bottles never see the market, because they are served at official government banquets. I have had the privilege of tasting the allocation reserved for the People's Liberation Army, which characteristically comes in a camouflage bottle. The highest grade of Wuliangye costs about $3,500. Newer baijiu variations include Vinn, which is produced in Oregon. Of the rice type, and aimed at the American market, it is one of the smoothest baijius I have tasted. Ming River is a strong variety from Luzhou Laojiao, a distillery that has some fermentation pits dating from 1573, and that is now venturing into the international market. Another baijiu manufacturer that claims a long history is from Shui Jing Fang, which is built atop a six-hundred-year-old distillery in Chengdu, Sichuan. Its product is of the strong classification and is aged for five years before being blended with spirits that are over forty years old. There are even antique varieties, some of which are liquors that were stored in basements that then were built over. One on my shelf (a gift from a friend) is called Daoguang. From a distillery in Jinzhou, it is in a 50 ml bottle that is dated 1845. It comes in a presentation case, with a book and a DVD about the "archaeological" site that it came from. I haven't yet tasted this one, saving it for my deathbed. Recent years have even seen the appearance of new artisanal baijiu brands catering to the explosion of a new middle class. And, for the impecunious, there is always Erguotou Red Star, for which a 100 ml bottle can be had for as little as $1.50.

Baijiu is consumed both formally and informally. It is a very social experience, and sometimes a physical test. I have had it poured from elegant individual crystal decanters, but I have also drunk it as a gift from a farmer who pulled out a small green Erguotou bottle from a patched, faded Mao jacket. At formal dinners, rare connoisseur bottles may be the centerpiece of a round table, with white tablecloth service. In more informal situations, it is often drunk by friends. An acquaintance with a very high-stress job used to consume a half-liter bottle several nights a week as he took a turn around Beijing's third-ring road. By pacing himself he could make the drive all the way around Beijing in the forty-two minutes and forty-nine seconds that it took to hear the entirety of Pink Floyd's *Dark Side of the Moon*. (That's impossible now, due to traffic and a crackdown on drinking and driving in the Chinese capital.)

Baijiu is not infrequently drunk in excess, since in China being drunk carries no stigma. Indeed, refusing a pour is a definite insult to the host, whether they are a rural farmer or a billion-dollar industrialist. You stop drinking only when the host stops. Generally, however, baijiu is drunk with meals, and I have seen many people carried out of fine restaurants after long drinking sessions. One of my more corpulent American colleagues had to be transported to his hotel room in a luggage trolley after such a bacchanal. Some hotels even offer rooms set up with intravenous fluids administered by medical personnel, to help guests recover more quickly.

Baijiu drinking accompanies most festive occasions, celebrations, and business transactions. At banquet meals there is a prescribed choreography and etiquette. Some of this depends on where you are drinking, when you are drinking, why you are drinking, and who is drinking with you. And while this is not a Japanese tea ceremony, there is a protocol. A few general rules are dictated by general Chinese Confucian tradition, and the three tenets of Chinese drinking culture: *Jiupin* (good manners), *Jiuliang* (high tolerance), and *Jiudan* (courage). The order in which people sit is dictated by who is the host, and if there is an honoree. The host and the most powerful guests sit at the head of the table, looking toward the door. Status declines to both the right and the left, down to the level of those who have their backs to the door and are often sent out on small errands (to get cigarettes, call for the check, or make sure the car is ready). Although men drink more heavily than women, there are plenty of formidable female baijiu drinkers. After the bottle is brought to the table and inspected by the host or a designee, it is poured into a pitcher made of a ceramic or, less traditionally, glass. Baijiu is always served neat and at room temperature, with individual aliquots usually dispensed from the pitcher, usually by a waiter or waitress, into small (about 15 ml) clear glasses. Just as tequila is measured in shots, and Scotch in drams, it is these small glasses, called cups, that are often the measure of baijiu, although any glass can be used in a pinch.

Toasting in Asia is related to the concept of face. Unlike in occidental culture, the first toast is always made by the host, usually to thank the guests for coming. You never take a sip until the host has raised his glass. Always grip the vessel with your right hand, and you can show greater deference to the host by resting it on the palm of your left hand. It is customary to completely drain your glass and proclaim "Ganbei" when you are finished (and perhaps turn over your glass to demonstrate a job well done). From here it can get really demanding, and in some cases downright dangerous. After the host has made the first toast or two, the individual toasts begin. These can be directed at your right or left, or at anyone around the table. Getting up and walking to the other side of the table is polite, and multiple people can be engaged. Often without preamble, the glasses are filled to their brims and downed in a single gulp. Often someone will be singled out to be assaulted with multiple toasts, one after another. This can go on for a long time, and no one leaves until the host (or one of his representatives) quits drinking, and all the liquor in the pitcher is consumed.

Baijiu drinking culture is alive and thriving in China, and I have had many memorable baijiu experiences in over thirty years of working as an academic and traveler there. In one case, a government official in Liaoning summoned a gigantic Manchurian woman, who returned with a basketful of what looked like old anarchist bombs but were in fact very old baijiu flasks, supposedly from before the civil war. These were sealed, and the only way of opening them was to use a hammer and a screwdriver to perforate them. It was winter in a windowless, rural, smoke-filled room with a flickering fluorescent light. Since the banquet was given in honor of our team's visit to the area, we were singled out. It got late, and my memory of events is happily nonexistent. At the end, I don't think I have ever felt so bad. Next morning, I decided that the only way I would feel better was to have a beer, and one of our crew, Xiao Hu (Little Tiger) kindly agreed to get a six-pack. She returned in a snap and dispatched the bottle cap. Thirstily I began to power it down, before realizing in a nanosecond that it was baijiu-flavored beer. I will spare you the sequel.

Chinese people today are flirting with a more Western diet and lifestyle, and expensive wines and microbrews have become freely available in major cities like Shanghai, Guangzhou, and Beijing. People are trying to be healthier, and the government has vigorously promoted cutting down on spitting, smoking, and excessive consumption of food and drink. On the flip side is the deep tradition in China of social drinking and hosting banquets, which has been enhanced by the improved living standards of many Chinese citizens. Throwing a big dinner, or going out with friends, is an important part of modern Chinese culture. And baijiu is often at the center of it all.

18

Grappa

Michele Fino and Michele Fontefrancesco

For many Italians, grappa used to say little about place, but a lot about people. It told of rooms full of cigarette smoke; of nights spent playing cards with friends; of strong and rough flavors and dizzying alcoholic levels; and of an anonymity that came from distilling the pomace of different grapes from distant places. All that was true until quite recently, but no longer. Today's memorable grappas come in stylish bottles bearing labels that describe a pomace, made from a precise selection of grapes, grown and distilled by companies with long, distinguished histories. A distinct color, evoking amber, or perhaps straw in summer, tells of many seasons spent aging in an oak barrel. The nose is well rounded, with hints of flowers and spice that make every sip enjoyable despite an ABV over 40 percent. Today grappa tells new stories about famous terroirs, and it adds prestige to illustrious wines, such as Barolo and Barbaresco, which fully express the complexity that lingers in their marcs. Grappa, in other words, has become a gateway for anyone who wants to discover Italy through its viticultural landscapes and unique fragrances. Before us were two such bottles, an embarrassment of riches.

All grappa drinkers remember their first sip: an immediate fruity note followed by a strong flavor that immediately fired up their throats with its scents and high alcohol content. Although mainstream grappa typically delivers 35 to 40 percent ABV, the alcohol level of this spirit can run much higher, especially in private domestic products. Grappa is a distillate typical of northern Italy, where it has been produced continuously since at least the fifteenth century. There this drink has served as a fundamental element of social life, a food supplement, and as an intimate part of what today we might call a traditional rural circular economy (more on this later). Today, however, grappa is drunk all across Italy, and there are excellent grappa producers in central and southern Italy for whom grappa-making is less a tradition and more a matter of quality production and marketing.

The name *grappa* testifies to the popular origins of the distillate, and to a complex history that was further complicated by the political differences that characterized Italy before its unification in the nineteenth century. The word became an official part of the Italian language only in the early twentieth century, when it appeared in a dictionary in 1905. Before then, *grappa* was used only in the northeastern part of the country, and in other regions the same distillate had different names. To people in the Veneto it was *graspa* or *grapa;* in Piedmont in the northwest it was *branda;* and to residents of Sardinia it was *fil' e ferru.*

Grappa is made by distilling the pomace (also called the marc) that is left over when grapes have been crushed to make wine: it consists of the grape skins and seeds, possibly accompanied by other must residues such as grape stems. The distillation of grappa aims at extracting desired compounds (such as water, ethyl

Figure 18.1. Copper alembics at Montanaro, Barolo's oldest grappa distillery.

alcohol, "good" acids, and valuable minor alcohols) by steaming (the steam is produced by heating the pomace, or is generated externally and blown into the still). Then unwanted components are eliminated by removing the "heads and tails": the initial fraction in which the methyl alcohol is concentrated, and the final one that contains most of the unpleasant aromatic compounds. The alcohol content is finalized by adding water to the distillate. In starting with the pomace, grappa thus differs from those wine spirits such as Cognac and the other brandies that result from the distillation of wine. It also differs from the grape brandies (a new kind of spirit invented in Italy in 1988 by the renowned grappa maker Nonino) that are made out of the distillation of not-yet-separated wine and marc (Figure 18.1).

Long ago, grappa was generally made by farmers using makeshift stills. After a much stronger pressing than is given today, the marc was left in the press, or buried, until winter and the first snow came. Winter brought free time to the farmers, and snow was used as the cooling medium needed for the condensation of the alcoholic vapors produced during distillation. Thanks to this practical approach, generations of farmers who produced their own wine were able to drain the last drop of alcohol out of the already well-squeezed pomace. They did not see much need to worry about the quality of the final spirit, a product that was often deeply affected by the oxidation of the raw materials, and tainted by generous hints of vinegar and other unpleasant aromatic notes. Domestic grappa was thus typically not at all refined; instead it was a full-blown distillate, often with a harsh organoleptic profile. Downing a grappa was considered proof of one's masculinity.

Alongside domestic grappa production, from the eighteenth century onward, commercial grappa distilleries popped up and increased in number. Today's professionally produced grappas are quite different from the homemade versions, featuring a more polished taste and, generally, a lower alcoholic content. What is more, buyers today can choose between mainstream, mass-market grappas, and higher-quality boutique products. Mainstream grappa is made with unselected pomace and in continuous stills. The distillate is physically treated to reduce its roughness, and aging is not always completed. In contrast, high-quality boutique grappa is typically made with a single-varietal pomace from a prestigious wine producer. The crush is distilled fresh by skilled master distillers who take care to preserve the complexity of a grape varietal, or even of a terroir, and thus offer a superb experience to the drinker. Because of the many differences in production methods, grappa is sold at a remarkably wide range of prices. In Italy mainstream grappas, marketed without reference to individual grape variety or geographic origin, can be found for less than ten euros a bottle. For a quality grappa, in contrast, the price can easily surpass eighty euros per liter, because of the added costs imposed by aging, the prestige of the distiller, the origins of the pomace, and the designation of origin of the wine from which the pomace derives.

Since February 13, 2008, the word *grappa* is no longer simply the traditional name of a marc distillate—it is an officially protected geographical indication (PGI) in the European Union. This means that the term can no longer be used except to indicate a product of the distillation of pomace in Italy. Oddly enough, before 2008, Italy had already requested and obtained E.U.-protected geographical indications for many regional grappas: Lombardy, Piedmont, Friuli, Veneto, Trentino, Alto Adige, and Sicily have all been protected since 1989. (The PGI Grappa della Valle d'Aosta is currently in the "applied for" status and, when granted, it will firmly establish northern Italy as a mosaic of regional PGIs devoted to grappa.)

How was it that grappa was protected as an Italy-wide PGI only after several regional grappas, as well as grappas linked to specific wines with a designation of origin such as Barolo and Marsala, had already been protected? The explanation lies thousands of miles away. Around the year 2000, an international trade and cooperation agreement between the European Union and the Republic of South Africa acknowledged the extensive use of the term *grappa* to designate distillates produced and widely sold in South Africa. First through diplomatic channels, and then through PGI protection, Italy worked to reserve this use of the word only for products made in its home territory. In fact, as a condition of approving the overall trade and cooperation treaty, Italy requested that South African producers stop using the word *grappa,* and resort to the generic "grape marc spirit." The situation has not yet been resolved, and South African producers still bottle spirits under the *grappa* name, although they cannot sell the products that way in Europe.

As for any other PGI or PDO (protected designation of origin) product in Europe, the specifications approved by the EU Commission determine how it can be made. The current language for PGI grappa entered into force in 2016, and it was recently updated with an Italian Ministerial Decree released on January 28, 2019. This allows grappa to be made with the marc of any Italian designation of origin and geographical indication; it regulates the techniques that can be used; it sets the limits for the addition of sugar (twenty grams per liter) and caramel (2 percent by volume, but only for aged grappas); and it establishes the rules for dilution with water and for labeling as aged.

The modern squabbles over the use of the designation *grappa* suggest that it is important to look at its meaning. The word may in fact ultimately be of Germanic origin, stemming from *grap,* an old term for a hook used to gather grapes—think "grapple." One can imagine a link between the sharp flavor of grappa and hook-like strength, or even that the strong taste of the drink drew parallels with the diabolical instruments carried by Krampus, the Central European folk daemon companion of Saint Nicholas who visits homes during the Christmas season to punish bad children. The modern use of the word is rooted in the part of Italy to which this beverage came first: the Po Valley, in northern Italy. *Grappa* as we know it today derives from the Lombard and Venetian dialects, in which the drink is called *grapa*. This term derives from *grapus* (Latin for grape) or *graspo* (Italian for the stem of the grape bunch that includes the woody part but excludes the berries). The stems remain after a grape bunch is processed during winemaking, and they are distilled along with the pomace to make the liquor.

The details of grappa's history have been to some extent mythologized. In the advertising of many distilleries, as well as of the Italian National Grappa Institute, the origins of the distillate are traced back to antiquity, more precisely to the late years of Ptolemaic Egypt. It is safer, however, to link the beginnings of grappa to late medieval times, when distillation was introduced to Italy from the Arab world (see Chapters 2 and 3). Distillation was initially an alchemical activity, conducted for pharmaceutical uses; only later, during the Renaissance, did grappa gain repute as a recreational alcoholic drink. The culture of pomace distillation rapidly took root in different parts of northern Italy, leading to the proliferation of local names for the product. Among the oldest companies specializing in grappa distillation, starting in the eighteenth century, is Nardini Distillery in Bassano del Grappa (close to Vicenza, in the Veneto region): it is the oldest operating distillery, still controlled by the same family.

Even as the early commercial distilleries began production, distillation spread widely in the rural areas of northern Italy. Indeed, the success of grappa can be measured by the sheer extent of this at-home, unlicensed peasant production. In a place where viticulture can be carried out virtually anywhere, and where winemaking played a major part in the rural household economy at least until the first half of the twentieth century, each farm would produce a small quantity of grapes to satisfy the family's need for wine. In this context, making grappa was an important step in completing the rural "circular economy" of households; that is, it helped to maximize the use of each farm's agricultural products and byproducts, and thereby to reduce waste almost to zero. Distillation used the leftovers of wine production to produce grappa, which was then put to use as a drink; an agent in food preservation; an ingredient for cooking; a popular cure for coughs, indigestion, and headaches; and an antiseptic for cuts and other injuries.

Grappa's importance, then, lay first and foremost in the lives of ordinary people. While bourgeois and noble Italian houses were well stocked with expensive and often imported bottles of whisky, Cognac, and *rosoli* (sweet liqueurs derived from the infusion of flowers and fruits in alcohol), it was cheap grappa that was ubiquitous in the houses of factory workers and farmers. Grappa also had a certain clandestine character. Starting in the nineteenth century, and despite severe controls imposed by the state, small distilleries spread widely through the countryside, and sometimes also found their way into urban centers. The spirits they made were hand-crafted by their owners, often local blacksmiths whose tools came in handy for making distillation equipment. To complete the entire cycle of production, a producer actually needed very little: just a table and a few square meters of floor space to mount the still, boiler, and condenser. That was enough to produce the few bottles intended to be consumed at home and among friends. Those domestic distilleries were mostly discreetly hidden in cellars or attics to escape the notice of the police and neighbors' prying eyes. For decades, then, grappa was the Italian equivalent of moonshine, and the oral history of the countryside abounds with stories of nocturnal and secret distillations—all of which reinforced the clandestine, rebellious, dangerous reputation of this unique spirit. For many, grappa's danger lay in the risks of being caught by the police in the act of unlicensed distilling. Even today, in Italy the minimum fine for illegal distilling is 7,500 euros, and one can be sentenced to up to six years in jail. In addition, home distillation often meant insufficient ventilation, with the attendant possibility of severe alcohol poisoning, as well as the potential for improper fractioning to eliminate all the methanol "heads," which could have serious consequences for whoever consumed the product.

Domestic grappa was typically rough and "masculine," largely because an imprecise and usually shortened fractionation often left in some of the tails, the final part of the distillate. Although the tails are usually not that physically harmful, they do contain persistent bitter, acid, or oily notes that make the distillate very unpleasant to the taste: so unpleasant, indeed, as to make drinking a particularly rough grappa a ritual test of manhood. Adults also commonly drank grappa in taverns and bars (premises mainly frequented by men, especially at night), as a cheap alternative to wine. It was men's preferred drink during long nights spent playing cards and chatting at the tables, and eventually it became a common addition to coffee, giving the hot drink an invigorating alcoholic kick that was particularly appreciated in the morning or at the end of a substantial meal.

Grappa is above all a convivial drink. At home as well as in bars, it was and is a drink to share with relatives and friends, often accompanied by bread, cheese, and salami. The same conviviality has also been linked to the feeling of comradeship experienced during military service, particularly during the years after World War I when grappa consumption officially caught on in the army. During the war the front lines ran across traditional grappa-producing regions, so the distillate was easily available to the soldiers, who eagerly consumed it. It became an official part of the daily allowance of each soldier, particularly those soldiers deployed in the cool Alpine regions. After the war ended the army continued to use grappa, notably to keep its soldiers warm during winter watches. As a result, for World War I veterans, and especially those who had served in the *Alpini,* grappa became a symbol of the hard days of comradeship in the trenches. For the next generation, too, the spirit evoked the transition from youth to adulthood that compulsory military service had marked.

The consumption of grappa in the army also introduced the spirit to those beyond northern Italy, as people drafted into the military from other regions of Italy encountered it and spread the word. While all of the twenty distilleries founded before 1900 and still operating today are located in the north (six in Piedmont, six in Veneto, six in Friuli-Venezia Giulia, one in Lombardy, and one in Trentino-Alto Adige), after World War I large distilleries were opened in Emilia-Romagna, Tuscany, and Lazio to match the growing demand for the distillate in those regions. Consumption of grappa today is mainly domestic, and the few export destinations are mostly in nearby Europe, although in recent years bottled grappa imports into the United States have been growing. Germany receives the lion's share (86 percent) of Italian grappa exports, both in bulk and in bottles.

Grappa, then, lies at the center of a network of relationships that links farms, taverns, trenches, and barracks. It bridges the distance between the rural and the urban world. History has portrayed it as a cheap, unsophisticated, masculine drink. But that is changing. Over the past few decades the harsh features of the beverage have been tamed thanks to more refined distillation techniques. The latest generation of grappas has opened a new chapter for this historic drink, offering the possibility for drawing in women, young adults, and sophisticates looking for a smoother taste profile.

Such developments notwithstanding, grappa consumption has trended both up and down in recent years. In Italy the practice of concluding each meal with a sip of grappa, or of lacing coffee with it, hardly survives nowadays: the alcohol is just too strong and concentrated, especially given modern rules for driving. Recent consumers have thus tended to avoid grappa, even as they continue to drink wine or beer during their meals. Over the ten years from 2008 to 2018 (the last year for which production data are available), the production of grappa plunged, from around 300,000 hectoliters to just over 212,000 hectoliters annually.

But even as the quantities produced have been decreasing, the aromatic qualities of the grappas available have been consistently improving, so much so that grappa has increasingly become both the subject of innovative and refined new bottlings, and a favorite of mixologists. These changes mark a new chapter in the life of this distillate. From its early days as a rough peasant drink that took the measure of the drinker's virility, on through its time as an epic refreshment for soldiers at war and a symbol of the national spirit, grappa has emerged as a protagonist of novel gastronomic combinations, a refined ingredient that is increasingly beloved by trendy bartenders around the world.

19

Orujo (and Pisco)

Sergio Almécija

Back in my hometown in Spain, my father is a regular at a bar called A Rúa. Every time I visit, we go there and have a feast. Pepe—the Galician owner—always offers us his house digestif after our copious "mariscada." We know it's the good stuff because it comes in an old, chilled plastic water bottle that he has refilled himself. The source of its contents? Ignorance is bliss. So far all Pepe has ever told us is that it comes from someone in his home region. It is, of course, the original (white) orujo: transparent to the eye; refreshing to the nose; dry, dense, and bittersweet in the mouth; and soul-warming to boot.

It's also, for me, a way to know whom I'm drinking with: I suspect I will never call anyone a close friend who can't appreciate the postprandial health benefits of a "chupito de orujo."

Bagaceira in Portugal, *marc* in France, *grappa* in Italy, *tsipouro* in Greece, or *rakia* in Bulgaria, the *Aguardiente de orujo*—or for short, *orujo, aguardiente,* or even *caña*—is *the* spirit of Spain. Like marc, grappa, and the other spirits mentioned, orujo is a pomace brandy, a liquor obtained from the distillation of the solid remains of grapes used to produce wine. In its original form this spirit is transparent, with an ABV of over 50 percent. The distillation of orujo has long been a part of the lives of vine growers in areas of northwestern Spain: Asturias, Cantabria, Castilla y León, and, especially, Galicia (the home of Albariño wine). This beautiful region is one of contrasts: rugged coastlines with sharp rocky cliffs as well as lush, green, bucolic inland areas that were first settled by Celtic and other pre-Roman human populations.

Inhabitants of these parts of Spain have been distilling orujo since at least the sixteenth century, although the first description of alembics for the distillation of orujo appeared only in the seventeenth century, after the Spanish Jesuit Miguel Agustí described how marc was distilled in neighboring France. Alchemists from different convents and religious congregations rapidly shared this knowledge, which spread so quickly along the Camino de Santiago that one may assume that the active distillation of orujo proliferated there around this time, too. In Galicia, besides its use in the production of different liquors, orujo is a common ingredient in the kitchen, where it is used in the preparation of pear and apple jams. Orujo is also a medical remedy against headaches, toothaches, coughs, and other unwelcome pains. Apparently, orujo rubs help cows produce milk. And drinking orujo is a daily routine for laborers and sailors—especially before a hard day of work. As the saying goes, "A drop of orujo warms the belly and makes people strong and brave."

Orujo may be ready to drink in its year of its production, or it may go through an aging process. In Galicia, orujos are produced from the *bagazo*—the wet skins, stems, and pips left over from wine production—and the quality of this base is crucial to the quality of the final product. Because distillation capacity is less than the volume of grape pomace that typically comes in at harvest time, surplus bagazo is often silaged, for a maximum of five months, in airtight containers that usually have a capacity of between about two hundred and a thousand kilos.

Distillers in Galicia traditionally use alembics in which the boiler component is heated directly, using either firewood or gas. As usual, the distillation process occurs in two stages: first the volatile elements of the pomace are turned into vapor, then this vapor is condensed into heads, hearts, and tails. The proportion of heads separated off depends on the quality of the initial bagazos, and typically distillation stops when the distillate leaving the cooling system reaches 45 to 50 percent ABV. The tails are discarded. To obtain the best orujo quality, it is necessary to control the temperature of the distillate (which should always be about 18 to 20° C when it leaves the system). In Galicia, orujo is rarely distilled more than once. The original distillate has more alcohol than the product that is sold on the market: the ABV is lowered by adding water that is colorless, tasteless, and relatively salt-free (salts would precipitate, making the orujo cloudy). This type of water is difficult to obtain directly from nature, so distilleries nowadays commonly use demineralized water.

The last steps in producing orujo are stabilization, filtration, and aging. If the orujo is refrigerated directly after the alcohol dilution it will become cloudy, as the heavier molecules come out of solution. To avoid this, the orujo is stabilized gradually, for a few minutes to a few hours, at temperatures varying between 2 and –20 degrees C. Finally, the orujo is filtered to eliminate any particles remaining in suspension, and to obtain the characteristic crystalline shine. Then, if desired, it may be aged in a barrel of pedunculate oak, the qualities of which will greatly influence its final characteristics.

The basic aguardiente de orujo is usually the starting point for the production of other spirits, the most popular being *aguardiente de hierbas,* which cannot contain more than a hundred grams of sugar per liter, and *licor de hierbas,* which must have at least a hundred grams of sugar per liter. Each spirit must use a minimum of three different herbs in its production. Although any plant suitable for human consumption used is allowed for this purpose, the traditional recipes include combinations of mint, chamomile, luisa grass, rosemary, oregano, thyme, cilantro, and cinnamon. The spirits themselves may be infused with the herbs during distillation, or by later maceration. *Licor de café* is a variation in which coffee beans are used instead of herbs. And in some areas of Galicia, the locals also produce *abofado* by using orujo to stop the fermentation of a wine while it is still sweet (before all the sugar is converted to alcohol), in a process analogous to the one used in Cognac to produce the popular Pineau des Charentes.

One of the most popular orujo derivatives is *queimada,* or "fire punch." During the ritualized mixing ceremony, which many say originated in pagan times, sugar, orange and lemon peel, coffee beans, and orujo are added to red wine, and the resulting potion is heated while being stirred with a big spoon. The proportions of the added ingredients are primarily up to the discretion of the *queimador.* Singing an incantation to the characteristic blue fire that results from lighting an orujo-and-sugar mixture and stirring it into the bowl is an essential part of this ritual; it is intended to scare away evil spirits, and according to some chroniclers it was started by the Celtic peoples (the "Gallaeci") who occupied modern Galicia and surrounding areas during the Iron Age (about 1300–700 BCE). Even if some argue that this "old ritual" dates only to the 1950s, it would be churlish to ignore the possibility that this magical recipe has been influenced by diverse traditions from the many Celtic, Roman, and Arabic peoples who have inhabited Iberia over its long history.

Pisco from Peru and Chile requires special mention here, because it is sometimes considered a variation of the Spanish orujo. In fact, however, the two spirits are quite different, even if the distillation of pisco in Peru may have begun at around the same time as that of orujo in Spain. Pisco is produced by distilling fermented grape juice directly, instead of distilling the solid remains of pressing the grape. Pisco is thus similar to Cognacs and Armagnacs, but typically without long aging in wood barrels.

When the conquistadors arrived in the New World, they brought calamity with them. But they also brought cattle, olive oil, and grapes. Initially, wine was shipped to the New World from Europe, and because of its scarcity was reserved for church use in the sacraments. But vineyards were quickly created and, according to the Peruvian historian Lorenzo Huertas, the production of Pisco was probably under way before the end of the sixteenth century. In the beginning pisco may have been used mainly to fortify wine (preventing the activity of acetic acid bacteria) rather than for direct consumption, but the spirit became a key part of the Peruvian identity. The grape juice used was fermented, distilled, and stored in the clay jars—also known as "piscos"—that were commonly used as a medium of exchange.

The spirit takes its name from the Peruvian port city of Pisco, which became its most important center of distribution. The town was founded by Álvaro De Ponce in 1572, and while its original name was Santa María Magdalena, the city later took the name of the Pisco valley in which it is situated. The original roots of the name *pisco* are likely from the Quechua word for "bird," so it is thus of pre-Hispanic origin.

Returning to Spain, the total volume of orujo currently produced in Galicia is hard to estimate because of the highly traditional nature of its production and distribution. Toward the end of the nineteenth century the distillation of orujo was actually banned, in the belief that the spirit had a toxic component that could be lethal. Years later, distillation was resumed legally in some areas of Galicia; and, starting in 1911, some cities issued orujo distillation permits with certain restrictions (such as limited access to critical components of the distillation apparatus, so that only specific distillations would be performed). This system was extended to the rest of Galicia in 1927, under the law called Régimen Especial de Destilación de Aguardientes. This legislation established a tax system, specified where the orujo could be distributed, and allowed the use of movable stills, which were in common use among the traditional *alambiqueiros* (or *poteiros*). These changes produced significant developments in orujo distillation.

In 1985 the Régimen Especial was replaced by more restrictive rules, which required the distilling of orujo in fixed distilleries, leading to the disappearance of portable alembics and brutally impacting the lives of the traditional *poteiros*. Predictably, the clandestine production of orujo increased. The most recent iteration of the law, passed in 2012, continues to insist on the use of permanent fixed installations, while also mandating the use of traditional methods (such as alembic and direct fire) and requiring aging for at least one year in oak barrels of less than five hundred liters (or for at least two years in bigger barrels of five hundred to a thousand liters). The current law focuses on producing a spirit that is both safe to drink and that exhibits high standards of quality (for example, a refined color).

But we live in a world of unintended consequences, and the same law also encourages the small-scale production of "traditional" orujos whose origin can only be guessed from a piece of paper stuck to a refilled plastic bottle.

If you ever are lucky enough to be in Spain and in need of an excellent digestif, finding a good, legitimate orujo will be easy. Ruavieja is a very popular brand that can be found in every Spanish supermarket, and almost every bar or restaurant. This orujo is inexpensive yet elegant, and it is available in a wide range of styles: classic, or with herbs, coffee, and even cream. The traditional Aguardiente de Orujo Ruavieja comes in a characteristic clay bottle that keeps the contents cool once out of the fridge and on the table (or on the go). El Afilador Aguardiente de Orujo is also a classic in Spain. The company has been selling bottled orujo since 1943 (it was actually the first company in Spain to do so). In El Afilador, spirit lovers will find a traditional clean, clear, bright, and warm orujo.

If you are looking for a more innovative take on orujo, check out the winery and distillery Mar de Frades. In addition to its more traditional versions of the spirit, it has created an exceptional "original" orujo, in which local herbs as well as fruits (like the mirabelle plum) are macerated. This creamy spirit, with its copper-amber color, will appeal to adventurous palates. And because it is distributed internationally, you may even cross paths with this distinctive, delicious orujo in a liquor store near you.

20

Moonshine

Rob DeSalle

For this tasting we had to wear bib overalls, just as Ian's moonshining Tennessee in-law, Great Uncle Granville, had done back in the day. Corn-cob pipe between his teeth, this thoughtful and engaging man would always say that too much of his product would kill you, but a little bit of it would cure anything. The spirit we were about to sample, many years later, sat invisible inside a shiny red metal can of the kind we associated with turpentine. "Moonshine Whiskey," said the label, which also promised 100 percent corn in the mash (estate-grown, no less), and distillation in traditional copper stills. We twisted off the metal cap and gingerly poured a clear liquid into our glasses. The nose was fairly restrained, with an edge of corn sweetness. Cautiously, we took our first sip. We were very pleasantly surprised by the smoothness and lack of burn that we encountered, at least until the finish, which lingered quite a while. This was a pretty sophisticated spirit, if a bit monodimensional. Tasting it made us wonder what a couple of years in new charred barrels might have done for it—even if right now we felt pretty good and didn't regret their absence. Suddenly, those overalls seemed superfluous.

Moonshine, mash liquor, mountain dew, hooch, homebrew, white lightning, white whiskey, choop, shiney. What's in a name? These Americanized names for moonshine are small change compared to the hundreds of worldwide equivalents. The etymology of just one of these variations, *moonshine,* is itself convoluted. Its first recorded use in the thirteenth century was literal—the light of the moon—while its first use in the context of spirits comes down to us in the form of legend.

In late eighteenth-century England, where this legend originated, alcohol was an important source of government revenue, just like it is most everywhere. Imported spirits were regulated by customs officers, and the rural county of Wiltshire was apparently a pretty easy place to hide contraband spirits, especially brandy from France. In particular, ponds were great places to stow the illegal spirits in kegs hidden beneath the water. To recover them, the locals would use long rakes. When caught in the act of retrieving barrels by customs agents on a moonlit night, some yokels claimed that they were raking the cheese emitted by the moon shining onto the pond. Knowing no better, the citified customs officers simply assumed that the Wiltshire farmers were a bit slow-witted and, as the legend goes, the booze was left alone. From this time on, the epithet "moonraker" was applied to any Wiltshire resident who used a rake to retrieve kegs of brandy hidden in ponds. Once those Wiltshire residents' descendants had crossed the Atlantic, the term inspired the use of *moonshine* to refer to any illegally produced spirit. The word was mostly used in America's Appalachian and Ozark regions, very appropriately because the business of transporting the stuff was carried on at night.

A less romantic second version relates the term *moonshiner* to people who worked the late shift in a second job. In the case of those legitimately employed, the word eventually morphed into *moonlighters,* whereas in the Appalachians the original form was retained for the folks who made the stuff that bootleggers shifted in the middle of the night. The term *moonshine* became cemented into American English from 1920 to 1933, when alcohol production and consumption were outlawed in the United States during Prohibition.

Eventually, *moonshine* came to be used internationally in the English language to refer to any illicit clear and unaged whisk(e)y (including local Irish or Scottish versions). In fact, nearly every country on the planet has its own version of moonshine, whatever its name. Here we will focus mostly on the American form of this interesting spirit, whose manufacture and consumption are as much about flaunting a disregard for taxes and the federal government, as they are about sharing a back-porch drink with friends and family.

The story of American moonshine goes back to just after the U.S. Revolutionary War. In 1789, with the ratification of the U.S. Constitution, the new federal government could begin levying its own taxes. Additional import taxes were out of the question because, as Alexander Hamilton pointed out, those taxes were already inordinately high. But money had to be raised somehow: as a result of war expenses, the central government alone was in debt to the tune of $79 million, an almost unimaginable sum at the time. A solution to the debt problem, then, was one of the first matters of business for the new federal government. In 1791 the first new tax ever levied in the United States was the Whiskey Tax, or as some temperance leaders called it, a "sin tax." This tax on spirits was a serious move, because at the time making whiskey was more than just a way for rogues to fill their spare time. Many farmers with a subpar yield of crops would turn to making corn mash and distilling it, because they could profit more from moonshine than from selling the raw materials. And they simply hated to see this part of their livelihood taken from them by the taxmen or "revenuers."

By the end of 1791, the act was ready to be applied. But the idea of the tax rankled settlers on the western edge of the new country, who felt that the tax was unfairly aimed at them. Those westerners (actually living in western Pennsylvania) were the major producers and consumers of spirits, which they often relied on for their livelihoods. For two years they protested and organized against the tax, until in 1794 the bad feelings boiled over into a full-scale revolt. The Whiskey Rebellion, as it is now known, began when those western Pennsylvanians marched on Pittsburgh. After a major battle (the Battle of Bower Hill), more insurrection (another march on Pittsburgh by citizens displeased with the tax), and a treasonous meeting (at what is now called Whiskey Point), the rebels gained momentum.

Predictably enough, those tax-hating westerners stirred the ire of George Washington, the first president of the United States, who gathered thirteen thousand troops to put down the uprising. Alas, the troops were too much for the rebels, who capitulated to the federal government. By the end of 1794 the Whiskey Rebellion, the first challenge to the authority of America's central government, had been beaten down. Thomas Jefferson would repeal the Whiskey Tax in 1802. But while this lessened the tax burden for the time being, whiskey was still vulnerable to excise taxes and regulation. The tax was reinstated from 1812 to 1816 to pay for the War of 1812, and then again in 1862 to pay for the Civil War. It was not repealed thereafter, which meant that moonshiners had to add dodging the law to their repertoire of skills.

Spirits in the United States have always been the target of temperance movements; and when the most significant one occurred in 1920, moonshiners became the major source of spirits in many places. From 1920 to 1933, when Prohibition was repealed, moonshiners enjoyed their greatest expansion. The demand for their product was so intense that some lost their way, making cheaper, less reliable products and watering them down to boot. But in 1933 the demand for illicit spirits collapsed, and moonshiners by and large didn't adapt well. Moonshine took a bit of a dive, and it went into a period of ill repute that has lasted to the present.

Although moonshine may have lost its luster for most of the twentieth century, Prohibition inspired much of the culture and lure of moonshining that we cherish today. Bootleggers, not to be confused with moonshiners, became legendary figures. The illicit distributors of the products of moonshiners, bootleggers are named for the horse-riding smugglers who would hide the contraband in their riding boots. The battles between bootleggers and revenuers, many of which occurred in the Appalachian region of the United States, became part of the lore of moonshine history.

Reconstruction after the Civil War was rough on the illegal distillation industry. Residents of the hill regions resented the idea of being singled out by the revenuers for simply making a living, and took pride in thumbing their noses at the government. Eventually even many nondistilling mountain residents started taking part in frequently elaborate, community-wide warning systems that turned out to be so effective that, in 1876, there were an estimated three thousand stills functioning in the eastern hills of the United States.

Paradoxically, it was the hard-drinking President Ulysses S. Grant who enlisted Green B. Raum to head the Internal Revenue Service with a mandate to eliminate moonshine. Raum was very nearly successful, reaching nearly three thousand convictions in 1877 for evading the alcohol tax. But moonshine was down, not out. Because it was vastly more profitable to distill alcohol outside the law, the appeal of moonshining and bootlegging turned many once-legal distillers to the illegal trade in the 1880s and 1890s. Not surprisingly, the federal government responded by increasing the number of federal convictions for tax evasion on liquor.

This pattern of rebounding, and then getting beaten back by the tax man, is actually very similar to the interactions of predators and prey. Perhaps the most famous example of this is the moose-wolf dynamic on Isle Royale in Michigan, shown in Figure 20.1. A declining wolf population lessens predation on moose, which leads to more moose. Once the moose population reaches a critical point, however, the wolf population increases because there are enough resources for the wolves to do well. And once the moose population dwindles to a point where

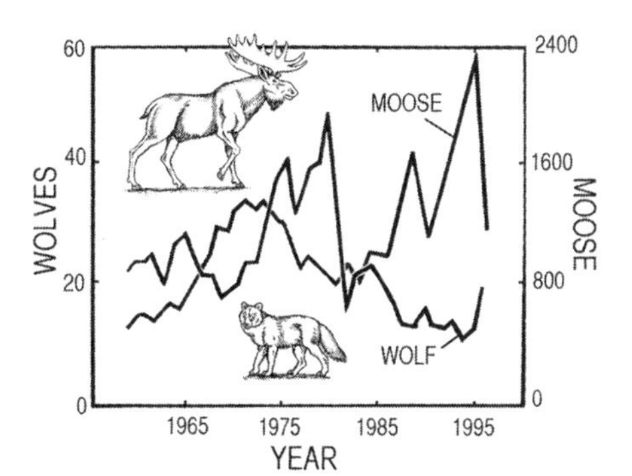

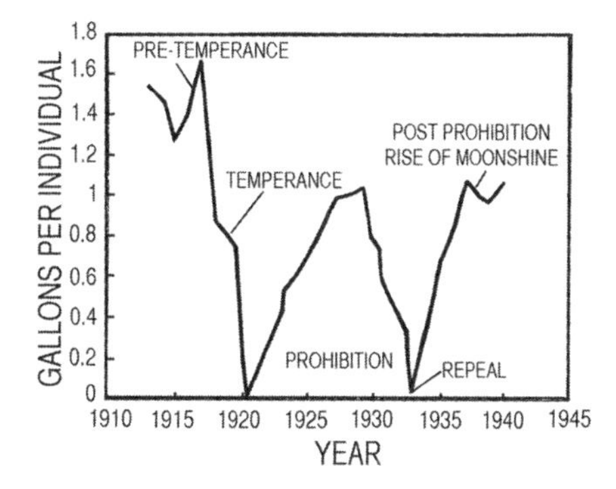

Figure 20.1. A predator-prey relationship in nature, and the cycle of alcohol use and social movements in the United States. *Left:* Moose and wolf population dynamics on Isle Royale in Michigan. The two populations appear to cycle as a result of their predator-prey relationships. *Right:* Gallons of alcohol consumed per individual from 1913 to 1940.

it can't support the increased wolf population, that population crashes, allowing for the moose population to increase. And so on. Each cycle from maximum population size to the next maximum population size takes about twenty-five years.

The cycles experienced by the wolf and the moose are very similar to those experienced by revenuers and moonshiners. Convictions for liquor tax evasion start out high. The convictions then drop as the moonshiners respond by going into hiding, or by dismantling their stills. The revenuers go away, and the distillers start back up again. Notice how the "predation" of the Internal Revenue Service on moonshiners as a result of the rise of temperance caused a large dip in moonshining that was evident from 1913 to 1922, shortly after Prohibition had begun. The extreme initial predation of Prohibition on its moonshiner "prey" pushed down moonshine production. But because there was such a big demand for alcohol during Prohibition, increased resources for the prey moonshiners allowed their numbers to rise. Enforcement pressure in the late 1920s caused yet another dip. Because alcohol was illegal there were no taxes on it, so that the official production curve dips all the way to 0.0 gallons per individual. But we know that moonshining was in its heyday during Prohibition, and our higher estimate for gallons per person comes from the fact that 1.1 million gallons of hooch were confiscated in 1929. There were about 110 million people in the United States at that time, which means that each individual averaged a tenth of a confiscated gallon that year. And while these calculations are very rough indeed, they do approximate the cycling of moonshine production in the United States from 1913 to 1940. Before 1929, which appears to have been the peak moment of Prohibition moonshine production, the federal government confiscated over a half million stills. After 1929 the confiscation of stills was also intense, so we show a peak for 1929 on the same level as the peak numbers for before and after Prohibition (Figure 20.1).

We could not find easily accessible data from 1940 to the present, but we do know that illegal moonshine had yet another peak in the 1950s, and a big dip thereafter. True moonshiners have pretty much died out today, although some in the hill regions of the United States still propagate the art. In contrast, the culture of moonshining has become increasingly mythologized in the public's eyes. Both movies and bluegrass music, in particular, have romanticized moonshiners and bootleggers. One 1950s movie, *Thunder Road,* was considered the iconic bootlegging movie, at least until *Lawless* came around in 2013. *Lawless* introduced us to the three Bondurant brothers, hell-bent on bootlegging hooch in the district of a corrupt sheriff and gangsters. The ebullient and overly handsome brothers are depicted as heroes. For its part, television usually presented moonshine in a humorous or even condescending light, though it also glorified producers like Jed Clampett on the *Beverly Hillbillies,* and the Duke brothers on the *Dukes of Hazzard,* while mocking law officers like Boss Hogg in the *Dukes* series. And moonshine bluegrass music is, simply put, a phenomenon.

Outrunning the local sheriff is the set-piece of any TV series or movie with moonshine in it. Bootleggers needed a lot of cunning, as well as tricked-out fast cars, to outrun the law and deliver the goods to the black market. In many ways, the NASCAR (National Association for Stock Car Auto Racing) phenomenon is tightly connected to bootlegging. Legend has it that the winner of the first NASCAR race had used his stock car to run hooch only a week before that initial 1949 race at North Carolina's Charlotte Raceway, which had been constructed by two bootlegging brothers, Pat and Harvey Charles. (The brothers were not present for that first NASCAR race because they had been convicted of bootlegging the year before.) The winning driver, Glenn Dunaway, won the two-hundred-lap race by three laps. But was the legend true? Well, when NASCAR officials measured the rear spring spread of his stock car, they found that the car's owner, Hubert Westmoreland, had spread the Ford's rear springs: the car was indeed souped-up using a common bootlegger tactic, and Westmoreland was a well-known bootlegger. You decide.

Moonshining has become something of a lost art. But there are still at least two moonshiners with what author Daniel S. Pierce calls "postmodern moonshine personas": Jim Tom Hedrick and Marvin "Popcorn" Sutton. Born in 1940 and 1946, respectively, Hedrick and Sutton (died in 2009) have facilitated the transition of moonshine from illegitimate to accepted and even legitimate. Both of these moonshiners are described as smart, homey guys whose collective knowledge of moonshine would put this chapter to shame. Hedrick, age seventy-nine, is still distilling and telling stories about it. He has teamed up with younger distillers of the Sugarland Distilling Company and is teaching them everything he knows. "Popcorn" Sutton, too, was a born storyteller and expert moonshiner. (Other former moonshiners, like Willie Clay Call who died in 2012, had also gone legitimate with their products; see Figure 20.2.). Hedrick and Sutton were both rather surprisingly religious. Pierce observes that when Popcorn Sutton was asked if he would do a photo shoot on a Sunday, he responded, "Hell no, god-

Figure 20.2. Legal moonshine.

dammit! If it's Sunday, I'll be in church." Another similarity is that both have been showcased on the Discovery Channel's runaway hit *Moonshiners.* If you have even a mild interest in moonshine, a few episodes of this show are a must-see. As with any reality series, some of it is set-up and hokey. But when Hedrick and Sutton were given free rein, they showed their originality and cleverness. They knew they were the last of a breed, and they were cashing in on it. Oh, and they always wore bib overalls. For that alone, they were classically cool.

If you started reading this chapter hoping that we would tell you how to make a still, we apologize: home distilling is illegal, at least for now. Furthermore, the best moonshiners may use rudimentary equipment, but they have broad technical knowledge about the process, which they use to keep themselves and their customers safe. In other words, good moonshiners are also good chemists. They have to be, not only to produce a drinkable product, but also to avoid killing themselves and others with it. That's because stills can be very dangerous. Under pressure they can explode; and because the product of a still is flammable, raging fire is an omnipresent danger, particularly after an explosion. In addition, if you don't know exactly what is in that innocuous-looking clear fluid coming off the still, you might get ethanol tainted with methanol—which, as you know from earlier chapters, is poisonous. It can cause blindness in small amounts, and in larger quantities is lethal. As a result, even in moonshining the making of spirits should be left to professionals. Thankfully, some of the products of these professionals are now legal, with unaged white whiskeys enjoying something of a renaissance, and a few even marketed as moonshine.

21

And for Good Measure . . .

Ian Tattersall

We knew that some wonderful wines come out of Lebanon. So why not an excellent grape spirit too? This one came in an unassuming bottle with a screw cap, but it glinted diamond-clear when the light caught it. Once in the glass, it immediately filled our noses with a distinct but not overwhelming aniseed fragrance that was backed up by hints of cedar and smoky caraway. From the very first sip its aromas lingered on our lips, and on the palate this arak was amazingly smooth despite its hefty 50 percent ABV. It yielded a mouth-filling warmth but no alcohol burn; instead that smoky anise flavor continued right through into a long, lingering finish. Poured over a cube of ice this arak took a while to turn cloudy, but the contact immediately softened its already understated alcoholic backbone while accentuating its anise flavors. The transformation revealed a strange and wonderful truth about this spirit: it has very distinct winter and summer versions.

An unavoidable problem for anyone writing a book about the astonishingly diverse world of spirits is that someone is bound to come along and complain that you have left out a favorite obscure tipple (ever heard of *starka*?). This book will sadly be no exception; we can hardly deny that there are spirits out there that clearly deserve mention but haven't received much if any attention so far. So, in a nod to comprehensiveness, this brief chapter will be devoted to some of these orphans. Not to all, alas, for that would be plainly impossible; but we encourage our readers to go out and discover for themselves as many of the absentees as they can. In that quest it is a good rule to keep an eye open for something you've never heard of whenever you find yourself in a liquor store. Because while not all spirits will be to everyone's taste, it's for good reason that so many of them have survived over the years. What's more, whether you personally decide you like it or not, at the very least anything you choose speaks to the spectacular cultural diversity that is the hallmark and glory of humankind. That's because, wherever they are made—which is almost everywhere—both the spirits themselves and the ways in which they are consumed reflect the essence of local life and tradition in their places of origin. Even in those parts of the world where the manufacture and consumption of spirits are legally forbidden, they are still enjoyed: one of my most treasured memories is of being smuggled into a speakeasy in Sana'a by a high official of the government of what was back then North Yemen. (My host ordered Scotch for us, though it could easily have been an arak.)

Given the ubiquity of obscure spirits, where to begin? Well, let's simplify matters by leaving aside those important and popular strong beverages, such as Japan's sakes, which, while comparable in strength to many apéritifs, do not strictly fit the spirits rubric. And let's content ourselves with the earlier brief mentions of colonial brandies and the Scandinavian tradition of mainly caraway- or dill seed–flavored aquavits (even if we did omit there extreme variants such as the Danish Gammel Dansk, which really lies in the bitters camp). That leaves us with three main categories of distilled beverages that deserve further mention: the many fruit-based spirits that we have not discussed as eaux-de-vie in Chapter 15; the grape-based araks of the eastern Mediterranean; and various southern and eastern Asian spirits, most notably the newly internationally fashionable sojus of Korea and shochus of Japan. And because spirits so readily invite mixing, it is also not too much of a stretch to include a brief allusion to the apéritifs and bitters that so frequently find their way into cocktails.

Let's start with those fruit spirits that we haven't considered under the eaux-de-vie rubric. Perhaps the best known of these come from central and eastern Europe, where the distilling of sugar-rich damson plums, in particular, has a long history. This activity produces beverages of around 40 percent ABV that normally bear some version of the name *slivovitz,* and that before World War II were often a staple of Ashkenazi communities, especially in Poland. In the northern parts of Croatia, along the border with Hungary, the spirit known as *Slavonska šljivovica* was traditionally once-distilled, in terracotta kettles, from the widely grown Bistrica plum and several indigenous varieties. After removal of the plum stones, the flesh was allowed to ferment naturally in wooden tubs for a couple of weeks before being boiled to separate the liquid. Nowadays small copper pot stills with a capacity of up to 150 liters are typically used for the distillation phase, and to receive the official Slavonska šljivovica designation the spirit then has to mature (for an unspecified period) in Slavonian oak barrels, from which it acquires a golden-yellow hue. The regulations also specify a "harmonious soft taste, rich aroma, and long finish," which is certainly true of the best examples we've tasted. Preferably served well chilled, this alluring spirit is indispensable in its home region as part of any special occasion.

North of the border, in Hungary, there is a long tradition of distilling that dates back to at least 1332. Its most renowned products in this linguistically distinctive region are the fruit brandies known collectively as *pálinka.* The law does not specify the kinds of fruit to be used for the beverages covered by this general name, but the plum spirits of the Szatmár and Békés regions have their own specific appellations. Six other Hungarian regions have similar local designations, each also tied to a particular fruit (apricots in Kecskemét and Gönc, for example, and sour cherries in Újfehértó). You are not likely to find many of these highly localized spirits in your neighborhood liquor store, but collectively they would make a great theme for a tour of Hungary. Be aware, though, that the pálinka designation means only that the contents of the bottle were fermented and distilled in Hungary from a mash of fresh (not dried) fruits that might have included concentrates and pulp. Flavor might also have been added by the introduction of dried fruits during the maturation process. What's more, while traditionally pálinkas were double-distilled in pot stills, at the lower end of the market you will nowadays find plenty of product that came off column stills before being macerated in steel or mulberry wood tanks. Confusingly, a bottle bearing the name *pálenka* might alternatively come from Slovakia and derive from almost any fruit, while *pálinka* is also legally produced in some regions of Austria.

When browsing in your local liquor store, you may occasionally spot a bottle with some version of the name *rakija* on the label. This is a blanket term for fruit spirits made in the Balkans and nearby; it would apply, for example, to that Croatian Slavonska šljivovica. Every country in this extensive part of the world energetically distills the fruits (and sometimes nuts, mainly walnuts) that grow there in such abundance. A wide variety of those fruits is distilled, ranging from apples and quinces through mulberries and peaches; and the terminology used for the results can vary confusingly. In Romania, for example, the term *rachiu* is applied to spirits derived from pears, apricots, and apples; but a distillate made from plums is a țuică. Whatever the name, such oddities are usually worth a try if you stumble across one. After all, distillers in these areas of Europe have been at their trade for a very long time.

Continuing to the far southern end of the Balkans, Greece has its own distilling tradition that very arguably goes back to classical times and that, both in terms of language and style, serves as a sort of bridge to its counterparts in those countries farther east whose activities were constrained by the influence of Islam. Greece's most famous spirit is ouzo, nowadays usually derived from commercial neutral spirits and flavored with various botanicals that mainly include anise, as well as varying proportions of fennel, coriander, cardamom, cinnamon, ginger, and nutmeg. Like the French pastis, which is also principally flavored with anise, ouzo famously goes cloudy when diluted with water. This is because the terpenes in the anise are soluble (and thus imperceptible to the eye) in any solution that contains above 30 percent ABV. But they come out of solution and form a visible whitish precipitate when the customary added water causes the alcohol concentration to fall below this level. (Any anise-containing spirit that does not do this was stabilized in advance to prevent it.) When in Greece you will probably begin your meal by complementing your *mezes* with a glass of ouzo, and you are just as likely to accompany your dessert with a glass of Metaxa, a spirit derived from grape brandy that includes various botanical additions and sweetish Muscat wines from the island of Samos.

But don't restrict your explorations of Greek distilled beverages to these two classics. Like every other wine-producing country, Greece produces spirits from the leftovers of winemaking, occasionally leavening the result with the addition of figs. *Tsipouro* is a (usually) unaged spirit that is twice distilled from grape pomace, and tradition holds that it has been made in Greek pot stills since at least the fourteenth century. Anise is often added, but when it isn't, it can be a remarkably soft and pure spirit that is an excellent apéritif, or can provide the ideal finish to a meal. Traditionally it is served in shot glasses, but a good tsipouro really deserves a broad tulip glass within which its aromas can expand. On the large island of Crete, tsipouro has a close relative called *tsikoudia* that is distilled only once. When mixed with a little honey this spirit is known as *rakomelo,* or *raki* for short, providing a linguistic connection between the Balkan rakias and the anise-flavored and grape-derived araks produced in the Levant.

Greece's traditional rival Turkey, historically a crucial nexus between East and West, produces anise-flavored rakis using a base similar to tsipouro, but sometimes starting with dried grapes (raisins) rather than fresh ones. Despite Koranic injunction, raki was widely (and mainly domestically) produced in the nineteenth century throughout the Ottoman Empire, and was served to the public in *meyhanes,* special establishments run by Albanians and Greeks. By the end of the century, more raki was drunk in Turkey than wine; and Mustafa Kemal Atatürk, the founder of the secular Turkish state, is said to have drunk half a liter daily.

The arak of the Levant is not to be confused with the cane-based Indonesian arrack we encountered at the beginning of Chapter 7. With a variety of modern derivatives, the term *arak* might be the most ancient of the names currently used for distilled spirits; and at some times, and in some places, it seems to have applied to spirits generically. As produced and consumed in the lands at the eastern end of the Mediterranean, arak is a clear white spirit produced from fermented grape must. As is the case with its close Turkish and Greek relatives raki and ouzo, arak is flavored with anise (and occasionally also with dates, figs, and other fruit). Like its kin, arak goes cloudy when water or ice is added, and it is similarly often consumed alongside the *mezze* that almost invariably precede any large meal. Both the water/ice and the food are wise precautions, because arak is normally sold at a minimum of 40 percent ABV and can be as strong as 63 percent or so (126 proof). The spirit is normally added to the ice, rather than the other way around, because if the ice is added later an undesired oily layer tends to form on the cold surface of the liquid, as oils from the anise solidify around the floating cubes. Although historically circumscribed by law and Islamic tradition, arak is nonetheless distilled widely in the Levant (in both pot and column stills). It is also made across northern Africa and as far south as the Sudan, where it is known as *araqui.*

Moving south and east, one very unusual spirit that demands mention is the *feni* that is unique to the formerly Portuguese enclave of Goa, in India. Surrounded by Hindu austerity, Goa has traditionally had a much more freewheeling attitude to alcohol, and it produces a creditable sweetish rum. But its most unusual contribution to the spirits scene is feni, recently recognized by the authorities as a "heritage brew." The classic feni is distilled from the flesh of the cashew fruit, thereby making use of a material that is usually discarded after the pendant nut has been removed. Traditionally, the fruits were foot-trodden before having a heavy weight placed on top to maximize extraction of the juice; the liquid was then fermented for up to a week in clay jars buried in the ground. Distillation originally took place in earthen pots known as *bhatti,* and while today copper is preferred for this stage of the process, the collector vessels, cooled by pouring cold water on them, are commonly still made of earthenware.

Feni is triple distilled, with the first run producing the suggestively named *urrack* that has an ABV of about 15 percent. Fresh juice is then added to the urrack for the second distillation, and more urrack is added to the result for the final pass. This takes the ABV of the finished alcohol to about 43 to 45 percent. (The product of the second distillation is sometimes sold more cheaply as *cazulo* with a slightly more modest ABV of 40 to 42 percent.) Production of feni is mostly by large numbers of independent small distillers, who bottle it in anything handy and sell it at local markets. Quality thus varies greatly, and very little is commercially produced. The cashew tree was introduced to Goa by the Portuguese in the sixteenth century, and there is some consequent uncertainty over the origin of this Goan distilling tradition. That's because the term *feni* also covers a palm wine–

based spirit that is distilled in south Goa in a similar way; and this beverage might (or might not) have been distilled prior to the arrival of the Portuguese.

Even farther to the east, there is plenty to interest the spirits aficionado in the countries of eastern Asia. We have devoted Chapter 17 to the long Chinese love affair with baijiu, while Chapter 3 looks briefly at the drinking culture of Japan; but we would be remiss not to mention Korea's pretty broadly defined national spirit, *soju.* Soju is made nowadays not only from a variety of cereals including barley and wheat as well as the traditional rice, but also from potatoes, tapioca root, and sweet potatoes. What's more, modern sojus may vary in ABV from about 17 to 53 percent. The Korean name is noncommittal, simply meaning "burned alcoholic liquor," and the practice of distilling originally came to Korea in a roundabout way, via invading thirteenth-century Mongols. It is uncertain whether those invaders had learned the techniques of distilling while invading Persia, or whether they had invented them on their own. But it may be significant that around the city of Gaegyeong, where the earliest Korean distilleries were established, (typically unflavored) soju is still sometimes called *arak-ju.*

The traditional grain-based soju was made by distilling a rice wine in a device that consisted of a two-level boiling pot, with a pipe at the top through which the condensate passed before trickling down and being collected. Such equipment produced a beverage of 35 percent ABV, which remained the standard until 1965 when a rice shortage caused the government to change the rules. At that point, soju began to be diluted to 30 percent ABV, from neutral spirits that were column-distilled from sweet potatoes or tapioca root; the addition of various flavorings and sweeteners soon followed. The restrictions on the use of rice were relaxed in 1999, and a handful of top-end brands have since returned to pot distillation, even as the middle and bottom ends of the market have both drifted toward more dilute column-distilled versions of the spirit. Since 2015, fruit-based sojus have become not only available in Korea but also very fashionable, especially among the young, to the point where soju as a category now dominates the Korean market for spirits. Soju has also been amazingly successful on the international scene, as a more viscous and slightly sweeter alternative to vodka. It is distributed in more than eighty countries, and in 2013 it edged out vodka as the world's best-selling spirit.

In Korea itself, drinking etiquette continues to a remarkable degree to mirror social hierarchy. Less prestigious individuals pour for more important and older ones (though a senior member of the group may pour for a guest), holding the jar or bottle in both hands as a sign of respect that the receiver is expected to reciprocate. Food is almost invariably consumed alongside soju, and the drink itself is normally presented in shot glasses for shooting rather than for sipping.

Soju's island cousin *shochu* has a five-hundred-year history in Japan, although it almost certainly came originally from the mainland. It is similarly made from a range of base ingredients (sweet potato, barley, and rice are the most commonly used). The best shochus are single-distilled, thereby maximally retaining the flavor qualities of the foundational ingredient, and they are typically diluted to an average ABV of 25 to 30 percent. They provide a wide range of distinctive flavor profiles, and at their best they offer amazing purities of taste. Although many shochus sold domestically in Japan are multiply distilled and are aimed at the lower segment of the market, those that are exported are more reliably high-end than the typical domestic shochu, and you can generally expect them to have a clarity of flavor commensurate with what you pay for them. Shochus are often drunk on the rocks, or they may be mixed with water (sometimes warm) or fresh fruit juice; but the best of them should be drunk straight, at room temperature. Use a thin-walled glass.

Finally, a brief word about apéritifs and bitters. In earlier pages we explored and celebrated the world's most popular postprandial digestifs (brandies, whiskies, grappas, and so forth). But we have neglected the spirits-based apéritifs that go so well before a meal, that are the mainstay of most cocktail parties, and that, let's be honest, are always a treat either on their own or on ice. They could be the subject of an entire book, but here are some basics.

The variety of apéritifs available is by now tremendous, which makes sense since the tradition of macerating botanicals in spirits extends back into the mists of history. As far back as we have any record of alcoholic drinks, we know that those beverages were regularly leavened with the addition of herbs, berries, and other flavorings, most of them added at first because alcohol turned out to be a great solvent for many of the botanical compounds that were believed, occasionally correctly, to have medicinal qualities. In fact, it is fair to say that the tradition of mixing drinks grew out of the much older monastic tradition of making medicines, as monks and then apothecaries yielded to innkeepers as purveyors of drinkable spirits (see Chapter 3). More than any other category of alcoholic beverage, it was spirits that invited mixing; and for many years recreational drinkers answered the call mainly by adding fruit and other flavorings right before consumption. The English, with their rum-producing colonies, eventually became particularly adept at this sort of mixing.

Still, in western Europe it was the French distiller Marie Brizard who, with huge success, marketed the first anisette in 1755 (allegedly she obtained the recipe from a West Indian man whom she had nursed through a life-threatening illness). Marie's creation was a flavored spirit rather than a complex apéritif, but it was eagerly devoured in the burgeoning coffee shops of the time and opened new vistas to drinkers. The category of *ratafias* subsequently blossomed as distillers began to add fruit to the herbs and spices that, because of their allegedly curative properties, had initially been the dominant botanicals. The age of liqueurs began to emerge. Sadly, the spirit of innovation ushered in by Marie's product was summarily snuffed out by the 1787 French Revolution, which put an end to everything that might have been considered a luxury (there was little use protesting that your fancy booze was a medicine).

But as we now know, this wasn't the end of the story. Right before the French Revolution broke out, the practice of pre-mixing wines, spirits, and a variety of botanicals in a bottle had been brought to Italy by the Turin distiller Antonio Carpano, soon to be celebrated as the inventor of what we now know as vermouth. Carpano's original 1786 concoction combined wine, spirits, bitter herbs, and spices, and it proved so popular that he quickly followed it with dozens of variations of his own, even as rival vermouth producers blossomed. The Italian "Torino" style now spearheads the market for the sweeter red vermouths, though not many, alas, are still made in Turin. The French in their turn are renowned for the quality of their pale, dry vermouths, following the lead of the herbalist Joseph Noilly, who made the first of them in Lyon during the early nineteenth-century upheavals of the Napoleonic period. But in apéritifs anything goes, and just remember when mixing your next Manhattan, Rob Roy, or martini, that in a vermouth color is not an entirely reliable indicator of sweetness.

The range of liqueurs and bottled apéritifs currently on the market is nothing short of mind-boggling, and there is no way to do justice to even a fraction of them here. What would a cocktail-lover's life be like without Campari? Lillet? Chartreuse? Drambuie? Grand Marnier? Bénédictine? Amaro? Fernet-Branca? Not to mention that every one of these is, each in its own idiosyncratic way, a lovely drink on its own. What you wouldn't want on its own, though, is a glass of pure bitters. After all, the most classic of these started out as serious therapeutics, and who among us doesn't have childhood memories of being forced to choke down spoonsful of disgusting medicines because "the worse it tastes, the better it is for you"? Antoine Amédée Peychaud, originator of the bitters that still bear his name, was a nineteenth-century New Orleans apothecary who trumpeted his invention as a cure for virtually every disease imaginable; and the still ubiquitous Angostura Bitters were originally devised by Johann Gottlieb Benjamin Siegert, a German surgeon who was seeking a tonic to pep up soldiers who were wilting in the Venezuelan heat. Alas, those bitters failed miserably in their original missions, although there are still many who swear by the curative powers of a shot of Fernet-Branca the morning after a hard night's drinking. (Just be advised that any "hair of the dog" anti-hangover effect it exerts may have more to do with the drink's counteraction of alcohol withdrawal than anything else.)

As a minuscule ingredient in cocktails, however, bitters really come into their own. Now available in a bewildering variety of flavors from an equally bewildering variety of sources, bitters, highly concentrated mixtures of botanicals in an alcohol base, are the ultimate exemplar of the dictum "less is more." Bitters enliven, and may even define, any cocktail containing them. Try to imagine a Manhattan without those two final dashes of Angostura—an impossible feat. And what would an Adonis be without that hint of orange bitters?

22

Cocktails and Mixed Drinks

Christian McKiernan and Rob DeSalle

Because of the pandemic, it was not possible for us to cozy up to a mixed drink at a busy bar. But we could settle for making one ourselves—or for trying a somewhat new offering, a ready-to-drink cocktail. We went with the tried-and-true Bloody Mary, partly because it was 11 a.m. on a Sunday, and partly because we thought it a better bet than a cocktail that might depend on fresher ingredients. We chilled a couple of pint glasses, placed an olive and a celery stalk in each glass along with some regular-sized ice cubes, and brought out what looked like a twelve-ounce beer can. It was gluten free, 10 percent ABV, and spicy—advertised as a whopping five peppers. Instead of the sound of shaking that we would have heard in a bar, we heard a psssst as the can opened—apparently this beverage was canned under pressure. It smelled like a Bloody Mary, poured like one, and left a coating on the glass when tilted: all signs of a drinkable tipple. On the palate it was indeed spicy, although not excessively so; it was perfectly enjoyable. We sipped our drinks slowly as we pondered the future of cocktails after the pandemic.

When discussing mixed drinks, we need first and foremost to distinguish between mixed drinks and cocktails. Thought they were the same thing? Well, not exactly. The first published mention of the word *cocktail* was in 1798, in the British *Morning Post and Gazetteer*. Unfortunately, the article did not describe the libation involved, though it mentioned that it was "vulgarly" (popularly) known as "ginger." Five years later, the word made its first printed American appearance, in the *Farmer's Cabinet,* a journal published in Amherst, New Hampshire. But once more it was used without explanation, and indeed the beverage involved may not even have been spirituous. Shortly thereafter, however, we get an earful from the *Balance and Columbian Repository,* a weekly publication of essays and editorials that addressed the ideas and news of the day for the benefit of residents of Hudson, New York. In the May 13, 1806, issue a reader asked, "What is a cocktail?" and received this classic answer, penned by the paper's publisher, Harry Croswell:

> As I make it a point, never to publish any thing (under my editorial head) but what I can explain, I shall not hesitate to gratify the curiosity of my inquisitive correspondent: Cock tail, then, is a stimulating liquor, composed of spirits of any kind, sugar, water, and bitters. It is vulgarly called a bittered sling, and is supposed to be an excellent electioneering potion, inasmuch as it renders the heart stout and bold, at the same time that it fuddles the head. It is said also, to be of great use to a democratic candidate: because, a person having swallowed a glass of it, is ready to swallow anything else.

The *Balance* was a Federalist publication, and the chance to get in a dig at the rival Democrat-Republicans was, we guess, just too much for its publisher to resist. But in any event, thanks to the Reverend Croswell cocktails are evermore defined as a mixture of spirits, water, sugar, and bitters. That means that a Screwdriver, while clearly a mixed drink, is not a cocktail. And slings, which were created well before cocktails came on the scene, differed from cocktails in not containing bitters. Okay, okay, you say, but what would a Singapore Sling be without that dash of bitters? All we can say in response is that language evolves.

Although Croswell's is the first printed American definition of "cocktail," the word itself is obviously older. Its convoluted and somewhat mysterious origin has been investigated by many notable scholars, including the humorist H. L. Mencken, author of both the book *The American Language* and the essay "How to Drink Like a Gentleman." According to Mencken, there are at least seven credible stories purporting to explain how the word *cocktail* originated. One of them calls out that the valves at the bottom of booze casks are called "cocks," and the dregs within the barrels are called "tailings." A second story suggests that the word derives from *cocktay,* a corruption of *coquetière,* the name of a drink served in New Orleans. A mythical Revolutionary War tavern owner named Betsy Flanagan is at the center of a third possibility that additionally involves a stolen chicken; and kudos is also given by some to a real barmaid named Catherine Hustler, who apparently used rooster tail feathers to stir and decorate the "gin mixture" she served in her Lewiston, New York, tavern. Alternatively, it's possible that the word came from Mexico, where drinks were garnished with the root of a plant called *cola de gallo* (tail of a cock). Or maybe the drink is named after an Aztec queen? We may never know for sure.

As scientists, however, we're inclined to defer to the solution proffered by the cocktail savant David Wondrich, who conducted an extensive investigation into the etymology of the word *cocktail* and eventually settled on a story involving the similarity between the appearance of a garnished cocktail drink and the tail of a horse that has been stimulated in a certain unseemly way. It seems that, until relatively recently, horse traders used a trick to make their older, more worn-out horses (and dogs) seem perkier. This stratagem was called "gingering," "feaguing," or "cocktailing," and it involved inserting some ginger into the anus of the poor horse, which would respond by sticking its tail straight up like a rooster's. Apparently, in horses (and dogs) the elevated tail is ordinarily a sign of youth, so cocktailing evidently proved a good strategy for profitably disposing of horses that had passed their sell-by dates. Here is what Mencken had to say about the practice in his book *The American Language:*

> When cocktails under that name became really popular in England, which was not until sometime after the establishment of the American bars, we had no doubt as to the derivation. To us, it was a short drink that cocked your tail using the same metaphor as to keep your tail up. If you exhibited a sporting dog of the setter type which tends to carry its tail low except in action, the show photographer would tell you to cock that dog's tail. . . . A cocktail was therefore what I suppose today would be called a pepper-upper.

Mencken's use of the term *pepper-upper* is probably a double-entendre referring to the fact that pepper was sometimes used as an alternative to ginger when feaguing horses.

As we discuss this momentous subject, we are reminded of an episode from Larry David's *Curb Your Enthusiasm,* in which a debate drags on over whether the Cobb salad was invented by Larry's friend Cliff Cobb's grandfather at the Drake Hotel in Chicago, or by the chef Bob Cobb at the Brown Derby Restaurant in Hollywood. The Bob Cobb choice appeared implausible, because the alliteration of the name seemed too good to be true. But at least for the purposes of the episode it turned out to be the right one—although the Brown Derby concerned turned out to have been in Albany, New York. The point here is that when you try to pin down the origin of a legend, you may find that you can be right and wrong at the same time. So, while we can't guarantee the accuracy of Wondrich's story, we are happy to go along with it.

The British were notorious in the eighteenth century for the mixed drinks that were served in their punch houses. But while those drinks were indeed mixed, they were not really cocktails as defined by the *Balance and Columbian Repository* in 1806. And because cocktails require both strong spirits and bitters, the British practice of mixing bitter medicinal herbs in wine (an ancient tradition dating back at least to predynastic Egypt) also falls outside the category. As a result, it seems to have fallen to the Brits' transatlantic descendants to invent the spirits-based cocktail. We have no idea exactly what the bitters were like that Croswell referred to in 1806, but by 1824 Angostura bitters had been invented in Venezuela, and by around 1830, Peychaud's bitters were being produced. Both brands are still going strong. Exactly who it was who mixed the very

Mint Julep.

(Use large bar glass.)

1 table-spoonful of white pulverized sugar.
2½ do. water, mix well with a spoon.

Take three or four sprigs of fresh mint, and press them well in the sugar and water, until the flavor of the mint is extracted; add one and a half wine-glass of Cognac brandy, and fill the glass with fine shaved ice, then draw out the sprigs of mint and insert them in the ice with the stems downward, so that the leaves will be above, in the shape of a bouquet; arrange berries, and small pieces of sliced orange on top in a tasty manner, dash with Jamaica rum, and sprinkle white sugar on top. Place a straw as represented in the cut, and you have a julep that is fit for an emperor.

Figure 22.1. Jerry Thomas's 1862 recipe for mint julep.

first cocktail, probably toward the end of the eighteenth century, will probably never be known for certain. But the New Orleans apothecary Antoine Amédée Peychaud, who certainly had the bitters at hand, is credited with inventing the oldest documented example, the Sazerac (a recipe that changed in the mid-nineteenth century from the original Cognac to today's rye whiskey due to the phylloxera plague).

The history of cocktails since 1806 includes some legendary names, but none surpasses that of Jeremiah "Jerry" Thomas, an immensely colorful character active in the mid-nineteenth century. Born on Long Island, and briefly a gold prospector in California, Thomas owned and worked in bars across the United States, dazzling saloon patrons with his larger-than-life and outrageously bejeweled persona, as well as with his energetic displays of cocktail-making expertise. Since his career began before the invention of the inherently exhibitionistic cocktail-shaker, Thomas became a master at the art of mixing cocktails by tossing them between containers held in opposite hands; and indeed, his signature drink was the Blue Blazer, which involved setting whiskey ablaze and passing an arc of flame back and forth in this way. It was Thomas who broke the bartenders' professional code of silence by writing *How to Mix Drinks; or, The Bon-Vivant's Companion.* Published in 1862, and often referred to as the "Bible of bartending," this book ran to several editions (the last, posthumous, edition of 1887 appeared as *Jerry Thomas' Bar-Tender's Guide,* still in print today), and it was enormously influential, even though many of his instructions are hard to follow. (Luckily for us, his recipe for mint julep, reproduced in Figure 22.1, is unusually straightforward, and is very similar to its modern counterpart.) Significantly, Thomas's book starts with about a hundred recipes for punches, forerunners of modern fruity mixed drinks. Most of his punch recipes originated in England—a fitting starting point for the essentially American cocktail.

Another hallowed name in the annals of the mixed drink is the slightly earlier Frederic Tudor (1783–1864). Tudor was not a mixologist, but he did help spirits enthusiasts in a big way: he figured out how to transport ice without melting, by packing it in sawdust. Where would any American bar be without this clever innovation by the world's "Ice King"? In addition to Thomas and the Ice King, the journalist Anna Archibald has proposed a Hall of Fame containing eight other bartenders (Figure 22.2). This gallery of heroes includes not only Don the Beachcomber and Trader Vic, both pioneers of tropical mixed drinks, but several clas-

Figure 22.2. Timeline for the Bartender Hall of Fame. Each bartender's name is above each icon, with a nickname or context immediately below. Also included, for some, is the name of a drink for which the bartender is famous.

sic bartenders like Ada Coleman, chief bartender at London's Savoy hotel for the first quarter of the twentieth century, and "Cocktail Bill" Boothby, prolific author and San Francisco's leading bartender at the time of the 1906 earthquake. Over the years, giants of mixology such as these have kept the art going, and transmitted their expertise via books, teaching, and demonstrations.

The cocktail culture boomed during Prohibition but waned somewhat in the latter half of the twentieth century. Two trends were mostly to blame. First, many partygoers turned toward drugs and away from alcohol, as a distinct drug culture developed. And the culture of mixed drinks itself declined, morphing from the imaginative use of fresh ingredients to prefabbed bottled mixes. Happily, this proved to be a passing phase. The youngest member of Archibald's hallowed group is a modern mixologist who gained fame while running the bar at the Rainbow Room in New York City. During the 1990s, Dale DeGroff played a leading role in resuscitating traditional cocktail culture. He reintroduced high standards in both ingredients and bartending methods, and he influenced an entire new generation of bartenders.

The cocktail and mixed drink bartenders we have mentioned so far are, and were, intuitive chemists in much the same sense as the moonshiners we discussed in Chapter 20. Mixologists, however, are also intuitive neurobiologists, because they have an inherent sense of the mechanisms of taste and, especially, smell (Chapter 8). How the tastes of bitters interact with ethanol and water is no trivial matter, and neither is how the fruity tastes of modern mixed drinks, and even of old-time punches, are manipulated to influence the senses. Perhaps a little science can help here.

A funny thing happens when water is added to ethanol. And to understand what happens, and why, we need to revisit those ethanol molecules. Each ethanol molecule has two ends. One end is hydrophobic and tries to avoid water, while the other end wants to react with other molecules, including water and other ethanols. If a beverage has less than 15 percent ABV, the ethanol is considered to be dissolved in the water. Because the ethanol in such solutions is so dilute, aromas can break their way out of the solution into what is called the "drink headspace." In contrast, above an ABV of 57 percent or so, water is so sparse in the beverage that it can be considered to be dissolved in the ethanol. At those high concentrations of ethanol, any aroma molecules in the beverage are more or less

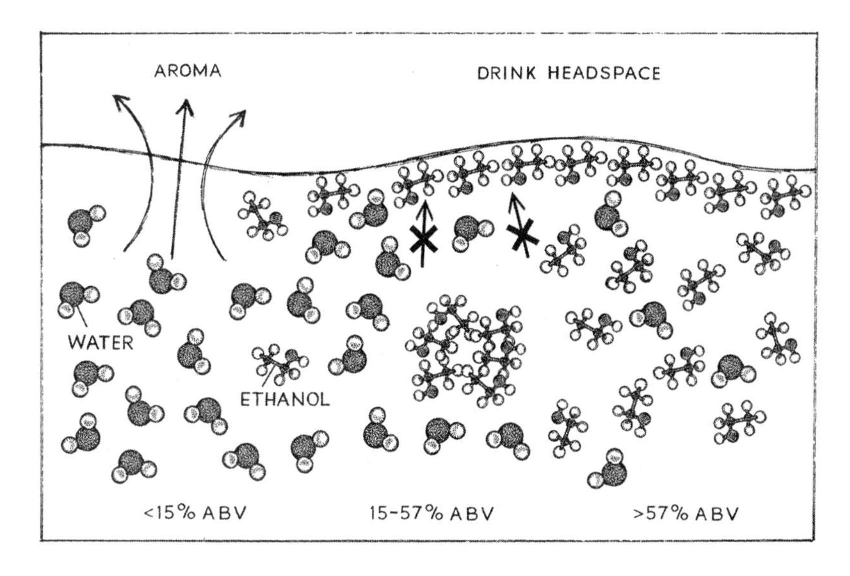

Figure 22.3. Diagram showing how and why aromas waft off spirits. Also depicted is how the aromas are held in solution until the concoction is drunk.

trapped beneath its surface, because the ethanol molecules will gather there and create a relatively impermeable surface layer. So as far as aromas are concerned, the "sweet spot" for most spirits is between 15 and 57 percent. In this range, the conditions of the solution are such that ethanol molecules will come together to form curious molecular structures called micelles. These small balls of ethanol molecules have a distinct structure, because the hydrophobic ends of the molecules burrow to the center of the micelle. The reactive ends of the ethanols thus stick out of the micelles and can react with other ethanol molecules and with odorants and other molecules that happen to be also swirling about in the solution (Figure 22.3). The ethanol concentration, then, affects how many of the odorant molecules in the solution can escape into the headspace where your nose can detect them.

This creates an interesting situation for the mixologist. Do you go easy on the alcohol, so that your drink will have a distinct and controllable aroma? Or do you push the limits with higher alcohol concentrations, and play with bitters or other additives that will appeal to the palate? Fortunately, there is a middle way: you can shake or stir. These mechanical operations break apart the ethanol molecules that are clustered on the surface, mixing everything up and allowing aroma molecules to get to the surface and waft out. Another good reason to shake or stir a mixed drink is that if you don't, the ethanol molecules, because of their chemistry, will cluster on the surface, making the concentration of ethanol higher at the surface and lower as you drink deeper into the cocktail—a phenomenon that no experienced drinker likes. Of course, as the mixed drink settles, the ethanol molecules drift right back to the surface; but for a brief time after shaking, the aroma molecules are liberated to waft up into the headspace.

This is where the neurobiology comes in. Remember that your sense of taste is almost as much about odorants as it is about the taste molecules you ingest. A simple exercise will help make this clear. Find a fruity-tasting jellybean or candy or, if you are in a bar, maybe a cocktail cherry. Hold your nose tightly, so you can't smell anything, and place the jellybean or cherry in your mouth and chew once or twice. You will get a little bit of the fruity taste from the gustatory molecules that are hitting your taste buds. Now let go of your nose and chew again. The taste should be more intense, and perhaps even be a little different from when you were holding your nose. This occurs because a sense-based phenomenon called "retro-nasal smelling" or "mouth smelling" kicks in. This is one of the main phenomena that flavor scientists study in order to understand the dual nature of taste.

One experiment by a group of flavor scientists compared swallowing an alcoholic beverage, as we normally do, versus spitting the drink out. This allowed the scientists to study the process of swallowing and explore the effects that it has on taste. In order to better understand the molecules released into the headspace and sensed that way, and those processed by swallowing, they co-opted some pretty advanced technology. Proton transfer reaction mass spectrometry was used to get a good picture of the aroma release in the nasal cavity. The results were then compared to findings obtained using "the temporal dominance of sensations method," which relies on subjects' own perceptions of odors and tastes, and is accurate to the extent that the subjects tested are trained to expertly detect certain odors and tastes. Using these approaches, swallowing a drink was shown to elicit more complex perceptions of taste than not swallowing did. This news will perhaps alarm the professional tasters of alcoholic beverages, who obviously have to spit if they are to taste dozens of specimens daily without getting drunk. But it seems that you really do have to swallow a drink to taste all of its complexity, something that will seem completely reasonable to the rest of us.

Another interesting study used "headspace–solid phase microextraction," along with gas chromatography–mass spectrometry, to measure the headspace aromas and their constituent molecular makeups. Researchers mixed up a mango-

and-vodka cocktail and with these tools found a very complex population of molecules in the headspace of this ostensibly simple mixed drink: it included no fewer than thirty-six volatile compounds (eighteen esters, ten terpenes, and eight other molecules). Limonene was found to be the strongest component, and the scientists used it in pairwise comparisons with other compounds that they found in high titers in the headspace of the drink. It turns out that both the molecular structures of compounds in the pairs (of which limonene was always one), and their ratios, affected the synergism of molecules in producing an aroma sensation. Molecules that are more similar in structure tend to be more synergistic with each other, and the synergistic impact is greatest when they are in almost equal ratios.

The reason we bring this up here is to illustrate the complexity and difficulty inherent in making mixed drinks that will elicit the best and most satisfying responses from our taste and olfactory systems. The simple combination of mango and vodka produces over thirty compounds, and as we explained in Chapter 8, our taste and olfactory systems need to process all of them for us to get the full experience the drink offers. Which brings us to the last topic of this chapter.

Virtual reality is all the rage nowadays; so much so that some researchers have attempted to explore the realm of virtual cocktails, or "vocktails." These virtual cocktails are not virtual reality in the strict sense of the term, but instead exist in the realm of augmented reality. Perhaps the closest thing to the vocktail is a virtual-reality game called *Season Traveller.* This game is played digitally, and it takes the gamer through four seasonal landscapes. The gamer is stimulated with wind, odor, and temperature changes, as well as with the audiovisual capacities of normal virtual reality. Vocktails alternatively use the device shown in Figure 22.4 to mimic sipping a complex drink. According to the creators of the vocktail, the idea is to create an augmented reality "that digitally simulates multisensory flavor experiences." In other words, the device creates digitally controlled taste, olfactory, and visual stimuli, and manipulates their interaction in order to simulate the sensation of drinking a cocktail.

The vocktail apparatus looks like a cocktail glass with a large cylindrical foot that houses its main mechanisms. The user adjusts the tastes, odors, and color desired in the vocktail to be "consumed," and an LED system illuminates the glass to give visual stimulation. The glass itself electrically stimulates the taste buds at the tip of the tongue to give the taste signals, and a series of tiny air pumps delivers odors to the imbiber. This all sounds incredible, but the device actually works quite well. Further, it has added to our understanding of the multimodality of

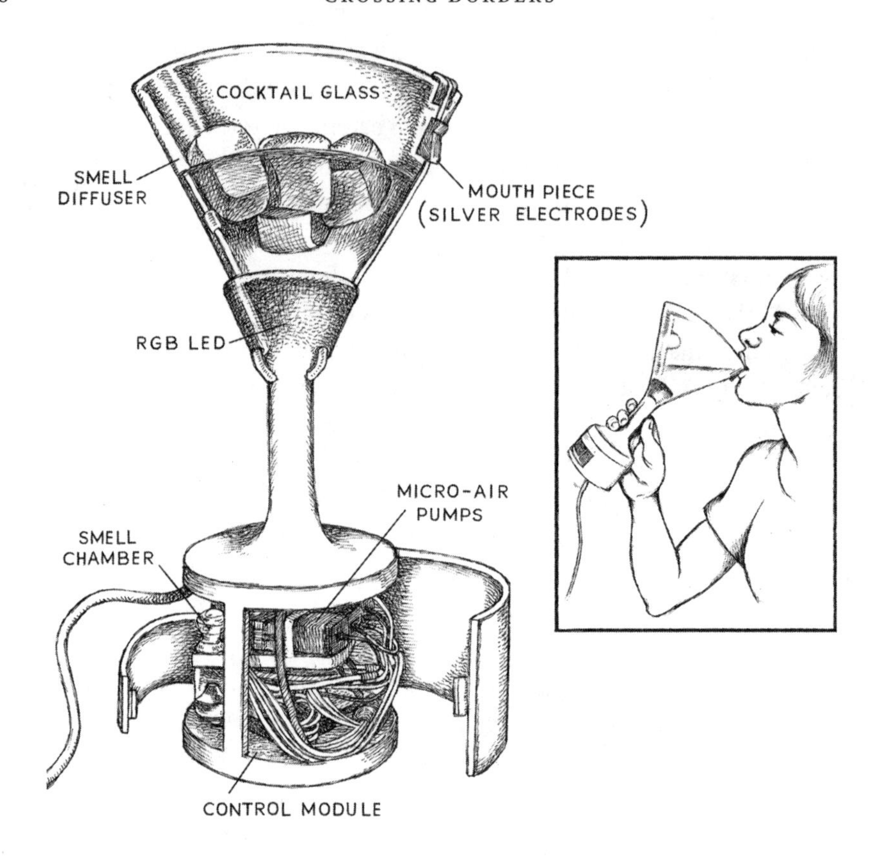

Figure 22.4. Structure and use of a vocktail device.

taste, olfaction, and sight. It can even be paired with a Bluetooth gadget whereby the "drinker" can adjust the sensations in real time. In comparative tests, the device accurately simulated both specific tastes, and combinations of tastes and odors.

So are we on the verge of virtual (and, incidentally, zero-calorie) cocktail bars? That seems unlikely, at least for the immediate future. The vocktail device was invented for investigating the multimodal interaction of sight, olfaction, and gustation when tasting sweet, salty, and sour things, and we are unaware of any plans to commercialize it. We suppose that one day the technology might be perfected to mimic the experience of drinking a menu of cocktails, and from there find its way into bars. But where is the buzz? The effect of alcohol is a much harder perceptual change to reproduce with augmented reality—as, indeed, is the delivery method. What is better than imbibing a real, viscid ambrosia from a cool, elegant, thin-walled glass, and feeling it move and expand from your lips to your throat? For now, then, our mixed drinks and cocktails are probably safe from augmented reality, although aficionados of nonalcoholic cocktails might cast envious sideways glances at that clunky vocktail apparatus. And as long as the new generation of mixologists continues to expand both the horizons of that movement and the range of gustatory experiences it offers, we are pretty sure that a virtual Moscow Mule is nothing more than a pipedream.

Which brings us finally to the more Generation X-ish endeavor of modern mixology. For while mixed drinks and classic mixology are old concepts, the roles of mixed drinks are changing among the newer generation. It remains true, of course, that virtually the entire template of mixology culture is the best part of two centuries old, coming as it did from the mind of Jerry Thomas, lionized as the "Father of American Mixology." In addition to the fizzes, punches, sours, and mint julep mentioned earlier, Thomas gave readers the very first written recipe of the Tom Collins, as well as of the Martinez, a concoction that is considered by some to be the precursor of one of the most iconic cocktails in history—the martini (this is actually a hotly contested claim). Thomas did more than write the book; he set the look and the lifestyle. He was the O.G. hipster of his time, who created the dress code of vest, pocket watch, sleeve garters, and other dandy favorites that became the uniform in hundreds of mixology establishments. Perhaps his unfortunate time as a minstrel-show manager helped the sense of theatricality he brought to the arc-of-flame bravura while concocting his signature Blue Blazer, but the performance art unquestionably became a thrilling component of a patron's experience.

Thomas's showmanship cemented his legendary status, and echoed time and time again throughout pop culture history. It undoubtedly begat the "flair bartending" that was seen on full and at times vomitory display in the 1988 movie *Cocktail.* In the film, Tom Cruise's neophyte mixologist Brian Flanagan, and his Obi-Wan mentor Doug Coughlin (Bryan Brown), dazzle five-layer-deep patrons at a series of saloons in New York City's Upper East Side. The customers don't seem to mind inordinate wait times for their drinks as long as they can watch the shakers and bottles juggled, tossed, flipped, and occasionally used to make cocktails, with almost no thought given to proper measurements. Product quality be damned, the movie was an international hit, and its influence certainly gives it a place in the pop-culture canon of American mixology. The movie offers no discernible recipes for a quality cocktail beyond the Red Eye, a hangover cure used by the bartending bros to bounce back from a night of "flairing." We cannot speak to the efficacy of their concoction, so if it induces gastrointestinal discomfort, listen to the Beach Boys' "Kokomo" from the movie's soundtrack to induce vomiting.

While flair bartending had its purpose (showmanship), the tastes and ideas of Millennials and Generation Z-ers seem to have expanded the approach to more multimodal creation of cocktails, and in the process have confirmed mixology as an art in itself. Modern-day mixologists spend their time at the boundary between practicality and creativity. An interesting but bad combination, or an excessively expensive drink, defeats the purpose of mixing drinks, which is to sell them. Avant-garde may be enticing, and have a funky look, but it simply isn't sustainable. Many master mixology courses focus on the practical and historical aspects of mixed drinks, and on how to improve the making of old drinks. This makes mixology distinct; and while at first mixology may not sound too different from bartending, call a mixologist a "bartender" and you will probably end up with flies in your martini. A bartender does not approach drink-mixing or even his or her patrons in the same way a mixologist does. The bartender is more practical, involved in running the bar and keeping the peace, while a mixologist is basically not interested in those jobs. A mixologist will instead know the history of cocktails, and how mixed drinks have evolved over time; and most importantly, he or she will try to contribute to that evolution. While a bartender will need to know how to make a defined list of cocktails with precision and efficiency, a mixologist, knowing the history of a cocktail, will play with how to reimagine it and change its appearance. A bartender has standard tools and ingredients that are easily obtained and mixed into drinks, while a mixologist will experiment with historical or other ingredients that are rarely used today. The difference is one of creative style: mixologists are the artists of drink mixing, while bartenders are the craftspeople.

Mixology is alive and well and will continue to flourish. But our changing social norms due to the COVID-19 pandemic might alter how we imagine mixed drinks and mixology in the future. How will drinking establishments survive the financial devastation caused by the pandemic? Philip Duff, director at Disruptive Craft Spirits, believes they'll get back to basics, with

> a return to simple drinks. Bars that *are* open are just desperately trying to survive, selling off their stock, doing takeout and delivery, and outdoor dining

> in commandeered parking spaces. This has led to a drastic drop in mixological snobbery; cocktail bars that wouldn't stock cranberry juice before because they didn't want to serve Cosmos are now doing strawberry margaritas and other simple, high-margin, crowd-pleasing drinks.

It remains to be seen if the spirits industry will adapt. Here again, Duff suggests that we'll be enjoying more cocktails in the can, and points out that canned cocktails are appearing more and more in liquor stores. The limited ability to go to a bar during the 2020 Covid pandemic and have an original and perfectly balanced cocktail handmade just for you could usher in a movement of high-end, ready-to-pour craft cocktails to be opened and enjoyed at home. Duff suggests that, as a result of COVID-19 and the closing of drinking establishments,

> home-based mixology has got, and will continue to get, the biggest shot in the arm since Prohibition, and it will be MEGA. For every million people baking sourdough or making banana bread during lockdown, at least 50,000 were freezing large-format ice, making sous-vide infusions, or upgrading their regular margaritas, manhattans and martinis with better liquor, liqueurs, vermouth, bitters and aforementioned ice.

All this could also open the way for high-end, ready-to-pour craft cocktails to be opened and enjoyed at home. If they are indeed high-end, which in general has yet to be proved, then the days of the booze-in-a-can stigma are numbered. Still, whatever the future trend in tippling, American mixology has not only survived but thrived over the past two centuries. We are clearly always up for another round of American ingenuity, flair, and creativity—think of all those dented shakers in home bars, and grenadine-stained rugs from failed attempts at "flair bartending." But please, let's not bring that back.

23

The Future of Spirits

Just as with the *fynbos* gin, we really had no idea what to expect from this "spirit whiskey with natural flavors." A radical departure from tradition, the product before us was the first "molecular" spirit, made to mimic an aged whiskey by analyzing which molecules give an aged bourbon its character, sourcing them separately, and adding them to a neutral grain spirit like those used for vodka and gin. We understood the impulse to shortcut the expensive, laborious process of barrel aging. But we did wonder how this futuristic marvel would taste. We were pleasantly surprised. Like a WWF wrestler, the squat cylindrical bottle with its short, stout neck subliminally promised us a powerful, even brutal, experience. But the fluid it contained was a light, golden straw in color, and the nose, though fragrant, was distinctly unaggressive, even retiring. This theme of understatement continued on the palate, which nonetheless yielded very distinct tones of vanilla and smoke, along with a hint of orange rind and cinnamon. The finish was soft, with minimal burn despite the 43 percent alcohol. And although we would personally have preferred something a bit more assertive and with a little more depth, we understood why the makers of this unconventional spirit had felt ready to launch their innovative product onto the market.

Spirituous beverages have come a long way in a thousand years, or even a hundred. Whatever your preferred spirit today, when you raise a glass you are unquestionably drinking a vastly superior product to whatever it was that the alchemists licked off their fingertips a millennium ago, or even to the firewater that most flappers downed in smoky speakeasies during the Roaring Twenties. Currently we are hearing a lot about how the post-millennial generation is turning away from strong spirits; and any consideration of where spirits are going has to take that into account. Yet we also have to acknowledge that recent statistics indicate that we are in a golden age for this rich and varied drinks category. Before the coronavirus crisis, the U.S. spirits market was predicted to grow by 6 percent annually, reaching an astronomical $58 billion by the end of 2022; and although alcohol sales stayed strong during the pandemic lockdown, it is hard to know how well pre-pandemic projections will hold up in the longer term. Nothing quite like this has happened in living memory, and how the bar scene will develop following the initial post-lockdown exuberance remains anyone's guess. Still, in the past spirits have proven quite recession-resistant. During the recession of 2007–2008, for instance, liquor sales—and profits—continued to grow, albeit at a reduced rate. This trend makes sense. Alcohol provides welcome solace in times of difficulty; and, at least by the shot, even premium brands of spirit are relatively inexpensive compared to most of the other small luxuries that help to sustain our morale during hard times.

There are nonetheless reasons to wonder whether recent trends will continue, at least for the immediate future. Much of the pre-pandemic increase in the value of spirits sales was driven less by volume than by consumers gravitating toward the more expensive super-premium and ultra-premium labels. In the eight years from 2012 to 2020, for example, sales of high-end spirits significantly outpaced the "value" market segment. But in a potentially weakening economy there is a significant risk that existing drinkers will reverse this trend by trading down instead of lessening their intake. And then, of course, there is the additional effect of the recent spate of politically motivated trade wars, which is already making itself particularly felt in the whisk(e)y business. Most notably, in 2018 bourbon from the United States found itself the subject of stiff retaliatory tariffs from the European Union, Canada, Mexico, and China, all important markets with enthusiastic consumers. "Trump's Trade War Leaves American Whiskey on the Rocks," blared a *New York Times* headline in February 2019. The article that followed left the reader in no doubt that "the trade war is stalling exports at a moment when thirst for products such as American whiskey is surging around the globe," and the author pointed out the disproportionate effects of this dismal development on smaller American distillers. As night follows day, American tariffs on Scotch and Irish whiskies ensued, upending the prospects of Celtic distillers who not only are producing blended spirits for export in the relatively near future, but also have to plan many years in advance for their barrel-aged products. Because of this unnecessary provocation, U.S. spirits exports fell 19 percent worldwide in 2019. And although American domestic sales boomed by 5.3 percent in the same year, clearly the market became skewed overall, creating painful uncertainties and producing particularly difficult conditions for the makers of bourbon and other spirits that require years of barrel aging. In the post-coronavirus world deglobalization seems likely to be a persistent theme, with the attendant risk that there will be fewer varieties of spirits to choose from at bars worldwide. If this happens, the craft distillers who survive the crisis will be most vulnerable: in an April 2020 survey, two-thirds of the distillers responding to an American Craft Spirits Association survey expected permanent closure if the pandemic conditions did not subside soon.

Well, that's the bad news, and only time will tell what long-term effects current political, epidemiological, and economic factors will eventually have on the spirits market. But from the individual consumer's point of view, drinking spirits is an aesthetic and gustatory matter rather than one of statistics, and on this more personal level the news is much more cheerful. For us consumers, the big domestic development in recent years has been a liberalization in regulatory regimes across the nation, with a loosening of restrictions that has encouraged an amazing florescence in the craft distilling industry. So even as the trend in industrial distilling has been toward the coalescence of large distilling enterprises into a small handful of behemoths, the diversity of individual products on the market has blossomed and the creativity of individual distillers has found unprecedented expression. From coast to coast, the choice of artisan whiskeys, gins, and vodkas has multiplied; and although prices have tended to reflect the diseconomies of small scale, discerning drinkers have found themselves presented with what amounts to an embarrassment of choice. Overall, the craft spirits market in the United States has expanded by around 19 percent every year since 2015. Very similar developments have also been taking place overseas, with gin producers in the United Kingdom, for example, multiplying sevenfold during the decade following 2010. Smaller producers are inevitably more adventurous and experimental than the established liquor behemoths, whose top concern is giving consumers a predictable product. So the most obvious result of the craft incursion has been an extraordinary selection of products for the discerning drinker to choose from, in virtually every category of spirits. Inevitably, this creative profusion will run into the tough realities of the marketplace, and over the next decade we will almost certainly see a lot of mergers taking place, as well as the disappearance of what once were highly promising labels. Already, Big Liquor is reacting not only by acquiring small producers, but also by coming up with new labels and new bottlings. Still, by now it is hard to imagine a future spirits industry without small, innovative producers at the fringes.

Consumer preferences drive innovation, and currently the millennial generation seems to be the most influential in shaping the market. Millennials already punch above their weight in the consumption of alcoholic beverages, comprising around 29 percent of the drinking-age population but accounting for 32 percent of the sales of all alcoholic drinks. Within the overall market, Millennials are tending away from beer and (to a lesser extent) wine, and toward spirits. In 2016 they drove the market share of spirits up to 36 percent of all alcoholic beverages by value, from 29 percent in 2000. Within the spirits category, whiskies, brandies (including Cognacs), and tequilas (especially the reposados) are gaining in market share among Millennials, at the expense of gin, rum, and even vodka (which continues nonetheless to rule the roost). Internationally, exotic spirits such as Korea's sojus are also restructuring the market.

This differential growth has taken place in the context of an increasing tendency to "drink in" at home, rather than to "drink out" at bars and restaurants—a trend that not only has implications for the hospitality industry and for the larger social scene in the age of the internet, but may eventually also have a knock-on effect on how spirits are packaged. Already, smaller and more portable packaging sizes are seeing greater market growth (from 2016 to 2017, sales of 50 ml bottles rose by 18 percent, while those of 100 ml bottles rose only 11 percent). As the vodka producers were inevitably the first to realize, packaging is a significant factor in this market. Cans have not yet made the headway in the spirits market that they have made in beer and even in wine, and when they do claim their space, they will probably contain mostly the premixed cocktails whose future looks bright if quality can be maintained. But the past success of the "miniature" single-serving bottle may also be showing the way forward for pure spirits.

What those bottles will mainly be made of is unpredictable, although a distinct move away from the traditional glass seems to be under way. In July 2020 Diageo, the largest player in the spirits business, announced that it would be moving its iconic Johnnie Walker whisky to bottles "made entirely from sustainably sourced wood [pulp]." Every year, spirits and other beverages are packaged in nightmarishly huge quantities of glass and—even worse—unrecyclable plastic containers. In an age in which consumers are increasingly aware of how our terrestrial and aquatic environments are being polluted by plastics (not to mention the ways in which they are personally being polluted by microplastics, as the environment takes its revenge), one can only hope that this will be the beginning of a significant trend. Diageo has formed a partnership with an outfit called Pilot Lite to produce wood-based bottles that will apparently also be used by PepsiCo and Unilever, and the Danish brewer Carlsberg has also promised paper-based bottles in the near future. This is a welcome move.

No discussion of Millennials' drinking habits can avoid the small but discernible tendency among members of this generation to turn away from all alcoholic beverages. At one time, the drinks market was fairly rigorously divided between its alcoholic and nonalcoholic segments, the former being somehow emblematic of "adulthood": in drinking families, even Grandma would take a glass of sherry from time to time. Now the lines are becoming blurred, and the mighty wellness-drinks industry has been explicitly intruding into the purview of the alcoholic sector with a highly imaginative variety of sophisticated, or at least exotically unfamiliar, beverages. These have found a receptive audience among Millennials, who are showing concern not only with issues of general health, but also with the alcohol abuse that is not hard to find among their parents. The makers of alcoholic beverages have predictably struck back. Brewers have done this most successfully, with a whole slew of nonalcoholic or low-alcohol beers that have lately vastly improved in quality, and that have been well received by consumers. Lowering the alcohol is not a strategy that the spirits sector can easily adopt (although low-alcohol sojus have been successful internationally, and the premixed cocktails that are trending today may achieve something similar); but a variation of that approach has led to the development of nonalcoholic "spirits" that mimic the gustatory qualities of the alcoholic originals.

The pathbreaker in this realm was the United Kingdom's Seedlip, now partly owned by the spirits giant Diageo (which has already invested significantly in the nonalcoholic sector). As we saw in Chapter 8, alcohol actually has a significant effect on the senses of taste and smell, and Seedlip started with a botanically complex "gin" that swapped out the juniper for botanicals specifically intended to help simulate the experience of drinking alcohol. Neutral spirits are used in the extraction phase, but they are later distilled out. The product has been well received in some influential quarters, and thanks to the Diageo connection, Seedlip is now quite widely distributed in three distinctive varieties (more are planned). Drunk straight, or even on the rocks, all three versions have received mixed reviews that range from the ecstatic to the defamatory (our own opinion leans negative). But in the hands of a talented bartender, any of them can undeniably form the basis of an interesting mocktail. An alternative approach has been taken by another British producer, Three Spirit, which blends botanicals with the intent of producing an effect on the brain and mind, rather than on the oral tissues—though in the end, the effect is perhaps not all that different. Conversely, Chicago's Ritual Whiskey Alternative specifically promises to deliver the complex "taste and burn" of whiskey without the alcohol, listing an impressive list of botanicals on its label. Ominously, though, that label also suggests that it is "best used in cocktails," which is excellent advice. Capsicum and green peppercorns certainly produce a "burn," but it is not quite the same burn one is looking for in a stand-alone spirit.

The bottom line at present, then, appears to be that nothing else yet on the market completely captures alcohol's mouth-filling bite. And although ethanol is far from the only mind-altering substance around, there is also nothing else currently available that gives that comforting warm buzz that you get from alcohol imbibed in moderation. But the science of nonalcoholic spirits is young, and who knows where human ingenuity will eventually lead? After all, the nonalcoholic, zero-calorie (and indeed, entirely immaterial) "vocktail" is already out there, at least in the laboratory (see Chapter 22).

Whatever the exact auguries for ersatz alcohol, it is safe to say that it will remain a specialist sector for the foreseeable future. By the same token, it is clear that our old, familiar alcoholic beverages are not going away anytime soon. Grappa is perhaps the traditional spirit that is currently most under threat. Its old-time devotees are aging and fading away, while the interests of many young Italians, living after all in a Nirvana of apéritifs, often lie elsewhere. But a devout grappa lover will optimistically explain that this is a transient marketing issue, rather than a problem with the spirit itself. And there is no question that in recent years grappa has been vigorously reinventing itself, not only by moving upmarket with products sourced from specific grape varieties, but also by offering bottlings from prestigious individual estates (see Chapter 18).

Even in the grain sector, change is likely; the craft grain distillers who currently buy their raw materials on the bulk market may well start along a similar path, as they identify particular sources of quality and reap the economic rewards of excellence. For in addition to the evident staying power of spirits in general (if 1920s bathtub gin didn't put drinkers off, nothing could), Millennials are a new audience, one interested in high-quality spirits, especially if they are made in an environmentally sustainable way.

Pernod Ricard, the world's second-largest wine-and-spirits group, has recently announced that, in keeping with U.N. ideals, over the next decade it will be implementing a new "sustainability and responsibility" program. There are four main goals, the first of which is "nurturing terroir." We have already suggested that most spirits lack terroir, as winemakers use the term, because they are conjured from diverse sources with the goal of making an extremely homogeneous and predictable product. Nonetheless, the company is explicitly acknowledging a responsibility to the land from which its raw ingredients are derived, and it has announced several pilot projects to protect local ecosystems and thereby assure the quality of its ingredients for years to come. Another area of attention is "circular making," a term that covers package recycling, reducing the carbon footprint, and the sustainable use of water. And Pernod Ricard is vowing to place more emphasis on valuing its employees, via strategies such as assuring gender balance, training workers to fit into a changing future, and cutting waste at points-of-sale. One intriguing initiative, for example, is to reduce the use of limes in bars by processing the used rinds into a concentrate instead of throwing them away. Finally, Pernod Ricard embraces "responsible hosting," by which it means actively working to stop the abuse of alcohol, especially among young adults. In doing all of this, the company's leadership is both responding to society's changing priorities and being savvy. For beyond the good public relations they engender, these changes reflect an understanding that a commercial operation can thrive only if it operates in a healthy society supported by a healthy environment.

Pernod Ricard is not alone among distillers in its new preoccupation. The Scotch Whisky Association has required all 101 of the malt and grain distilleries that comprise its membership to sign on to its Industry Environmental Strategy. This document mandates that members commit to reducing their use of water and packaging, and it requires them to aim for 20 percent renewable energy use by 2020. One way the distillers can achieve this is by using the "draff" (spent grain) from their stills as a source of heat (sometimes by converting it into methane gas), in much the same way that rum producers in the Caribbean and elsewhere generate electricity from burning the bagasse that is left over after crushing the sugarcane. Similarly, Sweden's Mackmyra Distillery has taken a page out of the winemakers' book, enhancing fuel efficiency by using gravity to move materials from one phase of production to the next, and by capturing all of the heat generated in the process. And in Mexico initiatives are under way to alleviate the environmental degradations caused by blue agave monoculture (Chapter 11).

But why go to all the fuss and bother of making your spirits using time-honored but labor-intensive methods, no matter how environmentally conscious? Traditional production methods are time-consuming, and the aging process in particular ties up capital for an inordinate time: surely there must be some way of bypassing such inefficiencies. Enter San Francisco's Endless West, maker of "molecular spirits." As we explained earlier, a chemist can characterize any beverage in terms of the molecules it contains. By one reckoning, for example, thirty different molecules account for the taste of a typical bourbon. In creating their Glyph whiskey equivalent, the chemists and culinary scientists at Endless West start with a "profile" for their desired whiskey that includes vital aspects of flavor, aroma, and mouthfeel. They determine which molecules are responsible for these gustatory features, then obtain them piecemeal from natural plant and yeast sources. In the laboratory, they combine these various elements in the predetermined proportions and add the result to a commercial neutral grain spirit. Once bottled, the Glyph is ready for immediate distribution—no wood or aging required.

Proudly advertised as "made overnight," this "biochemical equivalent" to "the finest aged whiskies" is said by its makers to have subtle aromas of vanilla on the nose; wood, spice, and hints of black fruit on the palate; and a "firm and earthy" finish. As we noted at the beginning of the chapter, this is a reasonably accurate description of Glyph's gustatory effects, and we think that the makers of this "spirit whiskey" have acquitted themselves pretty well, despite mixed reviews. There is clearly potential here, even if it may not have been quite fully realized yet. And let's remember that the folks at Endless West started with the hardest challenge first, since the farther any spirit departs from the neutral state, the more complex the mix of chemicals that compose it will be, and the harder that mix will be to replicate comprehensively. Chemically, American-style aged whiskeys are as complex as anything out there.

Human beings are above anything else social creatures, and in recent years the producers of spirits have increasingly come to appreciate this. Our image of the typical American whiskey drinker is no longer of a loner sitting in a gloomy, sticky-floored bar, silently knocking back shot after shot as the jukebox plays depressing country songs in the background. Nowadays, spirits are promoted as the ultimate social beverage, with glossy magazine ads showing expensive and prestigious labels being sipped by elegant drinkers of all ages—but particularly by the young.

You can plan to join fellow enthusiasts on organized whisky treks in places as far apart as Scotland and Tasmania, actively tasting and comparing a bewildering profusion of bottlings while enjoying wild nature to boot. If you're more sedentary, you can belong to a local society of enthusiasts who dispute at length about the shape and size of the glasses they use, and who employ an eyedropper to add distilled water for judicious dilution (actually, a single drop of water can momentarily smooth a rough spirit out amazingly well, probably by scattering hydrophobic molecules on its surface). Or you may prefer to open a bottle of your favorite spirit in your living room, in the company of friends. But whatever the context, and without denying that the occasional solo glass can cheer one up very nicely, it is usually the company that is as important as the whisky, ouzo, or brandy itself. The spirits are what bring people together—as likely as not, to argue about the relative merits of those very same spirits, or of whatever cocktail is preferred. It is the companionable nature of these magical beverages, whether mixed in an endless variety of cocktails, savored straight, or sipped on the rocks with a hint of soda, that is ultimately the most attractive thing about them. And while individual preferences will change, and fads will come and go, it is this sociable nature of spirits that banishes any doubt about their future staying power.

Further Reading

For those who wish to study spirits on a literary as well as a practical level, this bibliography is organized by chapter and so also by category of spirit. It contains all those works from which we have made citations, including some online references. These require the usual caveat that, while there is a great deal of information out there on the internet, its accuracy cannot be guaranteed.

Not sure where your interests lie? You might start with the excellent general information provided by Blue (2004), Kiple and Ornelas (2000), and Rogers (2014); Simon Difford's remarkable internet site, which is the most useful single source of information on both spirits and the world of mixed drinks; and *Whisky Advocate* magazine, which goes beyond the tight focus implied by its title to cover current developments on the general consumer-spirits scene.

Blue, Anthony Dias. 2004. *The Complete Book of Spirits: A Guide to Their History, Production, and Enjoyment.* New York: William Morrow.

Difford, Simon. "Difford's Guide for Discerning Drinkers." https://www.diffordsguide.com.

Kiple, Kenneth F., and Kriemhild C. Ornelas. 2000. *Cambridge World History of Food,* 2 vols. Cambridge, UK: Cambridge University Press.

Rogers, Adam. 2014. *Proof: The Science of Booze.* New York: Houghton Mifflin Harcourt.

Chapter 1. Why We Drink Spirits

Carrigan, M. A., O. Uryasev, C. B. Frye, B. L. Eckman, C. R. Myers, T. D. Hurley, and S. A. Benner. 2015. "Hominids Adapted to Metabolize Ethanol Long before Human-Directed Fermentation." *Proceedings of the National Academy of Sciences USA* 112 (2): 458–463.

Dietrich, L., E. Goetting-Martin, J. Herzog, P. Schmitt-Kopplin, et al. 2020. "Investigating the Function of the Pre-Pottery Neolithic Stone Troughs from Göbekli Tepe—An Integrated Approach." *Journal of Archaeological Science: Reports* 34, part A, 102618. doi.org/10.1016/j.jasrep.2020.102618.

Dietrich, Oliver, Manfred Heun, Jens Notroff, Klaus Schmidt, and Martin Zarnkow. 2012.

"The Role of Cult and Feasting in the Emergence of Neolithic Communities: New Evidence from Göbekli Tepe, Southeastern Turkey." *Antiquity* 86: 674–695.

Dudley, Robert. 2014. *The Drunken Monkey: Why We Drink and Abuse Alcohol.* Berkeley: University of California Press.

Hockings, Kimberley J., and Robin Dunbar, eds. 2019. *Alcohol and Humans: A Long and Social Affair.* Oxford, UK: Oxford University Press.

Hockings, Kimberley J., Nicola Bryson-Morrison, Susana Carvalho, Michiko Fujisawa, Tatyana Humle, William C. McGrew, Miho Nakamura, Gaku Ohashi, Yumi Yamanashi, Gen Yamakoshi, and Tetsuro Matsuzawa. 2015. "Tools to Tipple: Ethanol Ingestion by Wild Chimpanzees Using Leaf-Sponges." *Royal Society Open Science* 2: 150150.

McGovern, Patrick E. 2009. *Uncorking the Past: The Quest for Wine, Beer and Other Alcoholic Beverages.* Berkeley: University of California Press.

McGovern, Patrick E. 2019. "Uncorking the Past: Alcoholic Fermentation as Humankind's First Biotechnology." Pp. 81–92 in Kimberley J. Hockings and Robin Dunbar, eds., *Alcohol and Humans: A Long and Social Affair.* Oxford, UK: Oxford University Press.

Thomsen, Ruth, and Anja Zschoke. 2016. "Do Chimpanzees Like Alcohol?" *International Journal of Psychological Research* 9: 70–75.

Chapter 2. A Brief History of Distilling

Fairley, Thomas. 1907. "The Early History of Distillation." *Journal of the Institute of Brewing* 13 (6): 559–582.

Górak, Andrzej, and Eva Sorensen, eds. 2014. *Distillation: Fundamentals and Principles.* Amsterdam: Elsevier.

Hornsey, Ian. 2020. *A History of Distillation.* London: Royal Society of Chemistry.

Kockmann, Norbert. 2014. "History of Distillation." Pp. 1–43 in Andrzej Górak and Eva Sorensen, eds., *Distillation: Fundamentals and Principles.* Amsterdam: Elsevier.

McGovern, Patrick E. 2019. "Alcoholic Beverages as the Universal Medicine before Synthetics." Pp. 111–127 in *Chemistry's Role in Food Production and Sustainability: Past and Present.* Washington, DC: American Chemical Society.

McGovern, Patrick E., Fabien H. Toro, Gretchen R. Hall, et al. 2019. "Pre-Hispanic Distillation? A Biomolecular Archaeological Investigation." *Open Access Journal of Archaeology and Anthropology* 1 (2). doi: 10.33552/OAJAA.2019.01.000509.

Rasmussen, S. C. 2019. "From Aqua Vitae to E85: The History of Ethanol as a Fuel." *Substantia* 3 (2), suppl. 1: 43–55. doi: 10.13128/Sub-stantia-270.

Schreiner, Oswald. 1901. *History of the Art of Distillation and of Distilling Apparatus,* vol. 6. Milwaukee, WI: Pharmaceutical Review Publishing.

Webster, E. W. 1923. *Meteorologica.* Vol. 3 of *The Works of Aristotle Translated into English,* ed. W. D. Ross. Oxford, UK: Clarendon Press.

Chapter 3. Spirits, History, and Culture

Gately, Iain. 2008. *Drink: A Cultural History of Alcohol.* New York: Gotham Books.

Huang, H. T. 2000. *Biology and Biological Technology.* Part 5 of J. Needham: *Science and Civilization in China,* vol. 6. Cambridge, MA: Harvard University Press.

McGovern, Patrick E. 2017. *Ancient Brews Rediscovered and Re-Created.* New York: W.W. Norton. See esp. "What Next? A Cocktail from the New World, Anyone?," pp. 237–257.

McGovern, Patrick E., Fabien H. Toro, Gretchen R. Hall, et al. 2019. "Pre-Hispanic Distillation? A Biomolecular Archaeological Investigation." *Open Access Journal of Archaeology and Anthropology* 1 (2). doi: 10.33552/OAJAA.2019.01.000509.

Pierini, Marco. 2018. "The Origins of Alcoholic Distillation in the West: The Medical School of Salerno." Gotrum.com. http://www.gotrum.com/topics/marco-pierini.

Standage, Tom. 2005. *A History of the World in Six Glasses.* New York: Walker Publishing.

Chapter 4. The Ingredients

Biver, N., D. Bockelée-Morvan, R. Moreno, J. Crovisier, et al. 2015. "Ethyl Alcohol and Sugar in Comet C/2014 Q2." *Science Advances* 1 (9). doi: 10.1126/sciadv.1500863.

Charnley, S. B., M. E. Kress, A. G. G. M. Tielens, and T. J. Millar. 1995. "Interstellar Alcohols." *Astrophysical Journal* 448: 232–239.

Gottlieb, C. A., J. A. Ball, E. W. Gottlieb, and D. F. Dickinson. 1979. "Interstellar Methyl Alcohol." *Astrophysical Journal* 227: 422–432.

Kupferschmidt, Kai. 2014. "The Dangerous Professor." *Science* 343: 478–481.

Pomeranz, David. 2019. "The Inventor of Hangover-Free Synthetic Alcohol Has Already Tried It (and Hopes You Can Soon)." https://www.foodandwine.com/news/hangover-free-alcohol-alcarelle.

Qian, Qingli, Meng Cui, Jingjing Zhang, Junfeng Xiang, Jinliang Song, Guanying Yang, and Buxing Han. 2018. "Synthesis of Ethanol via a Reaction of Dimethyl Ether with CO2 and H2." *Green Chemistry* 20: 206–213.

Smith, David T. 2015. "The Fuss over Water." *Distiller Magazine,* July 2015. https://distilling.com/distillermagazine/the-fuss-over-water.

Wang, Chengtao, Jian Zhang, Gangqiang Qin, Liang Wang, Erik Zuidema, Qi Yang, Shanshan Dang, et al. 2020. "Direct Conversion of Syngas to Ethanol within Zeolite Crystals." *Chem* 6: 646–657.

Chapter 5. Distillation

DeSalle, Rob, and Ian Tattersall. 2012. *The Brain: Big Bangs, Behaviors, and Beliefs.* New Haven: Yale University Press.

DeSalle, Rob, and Ian Tattersall. 2019. *A Natural History of Beer.* New Haven: Yale University Press.

Górak, Andrzej, and Eva Sorensen. 2014. *Distillation: Fundamentals and Principles.* New York: Academic Press.

Hornsey, Ian. 2020. *A History of Distillation.* London: Royal Society of Chemistry.

Lane, Nick. 2003. *Oxygen: The Molecule That Made the World.* New York: Oxford University Press.

Moore, John T. 2011. *Chemistry for Dummies.* Hoboken, NJ: John Wiley & Sons.

Tattersall, Ian, and Rob DeSalle. 2015. *A Natural History of Wine.* New Haven: Yale University Press.

Winter, Arthur. 2005. *Organic Chemistry for Dummies.* New York: John Wiley & Sons.

Chapter 6. To Age or Not to Age?

BBC Scotland. 2018. "Third of Rare Scotch Whiskies Tested Found to Be Fake." https://www.bbc.com/news/uk-scotland-scotland-business-46566703.

Canas, Sara. 2017. "Phenolic Composition and Related Properties of Aged Wine Spirits: Influence of Barrel Characteristics: A Review." *Beverages* 3: 55. doi: 10.3390/beverages3040055.

Chatonnet, Pascal, and Denis Dubourdieu. 1998. "Comparative Study of the Characteristics of American White Oak (*Quercus alba*) and European Oak (*Quercus petraea* and *Q. robur*) for Production of Barrels Used in Barrel Aging of Wines." *American Journal of Enology and Viticulture* 49: 79–85.

De Rosso, Mirko, Davide Cancian, Annarita Panighel, Antonio Dalla Vedova, and Riccardo Flamini. 2009. "Chemical Compounds Released from Five Different Woods Used to Make Barrels for Aging Wines and Spirits: Volatile Compounds and Polyphenols." *Wood Science and Technology* 43: 375–385. doi.org/10.1007/s00226-008-0211-8.

Goode, Jamie. 2014. *The Science of Wine: From Vine to Glass,* 2nd ed., chap. 12. Berkeley: University of California Press.

Chapter 7. Spirits Trees

Fellows, Peter. 1992. *Small-Scale Food Processing.* London: Intermediate Technology Publications.

Goloboff, Pablo A., James S. Farris, and Kevin C. Nixon. 2008. "TNT, a Free Program for Phylogenetic Analysis." *Cladistics* 24 (5): 774–786.

"Know Your Whiskey." OldFashionedTraveler.com. http://oldfashionedtraveler.com/know-your-whiskey.

Li, Zheng, and Kenneth S. Suslick. 2018. "A Hand-Held Optoelectronic Nose for the Identification of Liquors." *ACS Sensors* 3 (17): 121–127.

Needham, Joseph P., with the collaboration of Ho Ping-Yü, and Lu Gwei-jin. 1970. *Science and Civilisation in China,* ed. Nathan Sivin, vol. 5, pt. 4: *Spagyrical Discovery and Invention: Apparatus, Theories, and Gifts.* Cambridge, UK: Cambridge University Press.

Rassiccia, Chris. "Whiskey Infographic," RassicciaCreative.com. http://www.rassicciacreative.com/tree.html.

Swofford, David L. 1993. "PAUP: Phylogenetic Analysis Using Parsimony." *Mac Version 3.1.1.* (Computer program and manual.)

Chapter 8. Spirits and Your Senses

Abernathy, Kenneth, L. Judson Chandler, and John J. Woodward. 2010. "Alcohol and the Prefrontal Cortex." *International Review of Neurobiology* 91: 289–320.

Abrahao, Karina P., Armando G. Salinas, and David M. Lovinger. 2017. "Alcohol and the Brain: Neuronal Molecular Targets, Synapses, and Circuits." *Neuron* 96 (6): 1223–1238.

Bloch, Natasha I. 2016. "The Evolution of Opsins and Color Vision: Connecting Genotype to a Complex Phenotype." *Acta Biológica Colombiana* 21: 481–494.

Bojar, Daniel. 2018. "The Spirit Within: The Effect of Ethanol on Drink Perception." https://medium.com/@daniel_24692/the-spirit-within-the-effect-of-ethanol-on-drink-perception-a4694c12322e.

Jeleń, Henryk H., Małgorzata Majcher, and Artur Szwengiel. 2019. "Key Odorants in Peated Malt Whisky and Its Differentiation from Other Whisky Types Using Profiling of Flavor and Volatile Compounds." *LWT-Food Science and Technology* 107: 56–63.

Lindsey, B. 2020. "Bottle Typing/Diagnostic Shapes." https://sha.org/bottle/liquor.htm#Bottle%20Typing%20Organization%20and%20Structure%20block.

Lumpkin, Ellen A., Kara L. Marshall, and Aislyn M. Nelson. 2010. "The Cell Biology of Touch." *Journal of Cell Biology* 191 (2): 237–248.

Plutowska, Beata, and Waldemar Wardencki. 2008. "Application of Gas Chromatography–Olfactometry (GC-O) in Analysis and Quality Assessment of Alcoholic Beverages—A Review." *Food Chemistry* 107 (1): 449–463.

Science Buddies. 2012. "Super-Tasting Science: Find Out If You're a 'Supertaster'!" *Scientific American,* December 27. https://www.scientificamerican.com/article/super-tasting-science-find-out-if-youre-a-supertaster.

Chapter 9. Brandy

Asher, Gerald. 2012. "Armagnac: The Spirit of d'Artagnan." In Asher, *A Carafe of Red.* Berkeley: University of California Press.

Cullen, L. M. 1998. *The Brandy Trade under the Ancien Régime: Regional Specialisation in the Charente.* Cambridge, UK: Cambridge University Press.

Faith, Nicholas. 2016. *Cognac: The Story of the World's Greatest Brandy.* Oxford, UK: Infinite Ideas.

Girard, Eudes. 2016. "Le cognac: Entre identité nationale et produit de la mondialisation." *Cybergeo: European Journal of Geography.* http://journals.openedition.org/cybergeo/27595.

Smart, Josephine. 2004. "Globalization and Modernity: A Case Study of Cognac Consumption in Hong Kong." *Anthropologica* 46(2): 219–229.

Chapter 10. Vodka

Herlihy, Patricia. 2012. *Vodka: A Global History.* London: Reaktion Books.

Himelstein, Linda. 2009. *The King of Vodka: The Story of Pyotr Smirnov and the Upheaval of an Empire.* New York: HarperCollins.

Hu, Naiping, Daniel Wu, Kelly Cross, Sergey Burikov, Tatiana Dolenko, Svetlana Patsaeva, and Dale W. Schaefer. 2010. "Structurability: A Collective Measure of Structural Differences in Vodkas." *Journal of Agricultural and Food Chemistry* 58: 7394–7402.

Matus, Victorino. 2014. *Vodka: How a Colorless, Odorless, Flavorless Spirit Conquered America.* Guilford, CT: Lyons Press.

Rohsenow, Damaris J., Jonathan Howland, J. T. Arnedt, Alissa B. Almeida, and Jacey Greece. 2009. "Intoxication with Bourbon versus Vodka: Effects on Hangover, Sleep, and Next-Day Neurocognitive Performance in Young Adults." *Alcoholism Clinical and Experimental Research* 34 (3): 509–518.

Shiltsev, Vladimir. 2019. "Dmitri Mendeleev and the Science of Vodka." *Physics Today,* August 22. doi: 10.1063/pt.6.4.20190822a.

Chapter 11. Tequila (and Mezcal)

Chadwick, Ian. 2021. "Tequila: In Search of the Blue Agave." IanChadwick.com. http://www.ianchadwick.com/tequila.

Martineau, Chantal. 2015. *How the Gringos Stole Tequila.* Chicago: Chicago Review Press.

Menuez, Douglas. 2005. *Heaven, Earth, Tequila: Un viaje del corazon de Mexico.* San Diego: Waterside.

Ruy-Sánchez, Alberto, and Margarita de Orellana, eds. 2004. *Tequila.* Washington, DC: Smithsonian Books.

Tequila Regulatory Council. 2021. https://www.crt.org.mx/index.php/en.

Valenzuela-Zapata, Ana, and Gary Paul Nabhan. 2003. *¡Tequila! A Natural and Cultural History.* Tucson: University of Arizona Press.

Chapter 12. Whisk(e)y

Broom, Dave. 2014. *The World Atlas of Whisky,* 2nd ed. London: Mitchell Beazley.
Greene, Heather. 2014. *Whisk(e)y Distilled.* New York: Avery.
Murray, Jim. 1997. *Jim Murray's Complete Book of Whisky.* London: Carlton Books.
Owens, Bill, and Alan Dikty. 2009. *The Art of Distilling Whiskey and Other Spirits.* Beverly, MA: Quarry Books.

Chapter 13. Gin (and Genever)

Anderson, Paul Bunyan. 1939. "Bernard Mandeville on Gin." *Publications of the Modern Language Associations of America* 54 (3): 775–784.
Broom, Dave. 2015. *Gin: The Manual.* London: Mitchell Beazley.
"Budget Blow Will Mean Price Hikes for Wine." 2018. WSTA.co.uk, October 23. https://www.wsta.co.uk/archives/press-release/budget-blow-will-mean-price-hikes-for-wine.
Stewart, Amy. 2913. *The Drunken Botanist.* New York: Algonquin.
Van Schoonenberghe, Eric. 1999. "Genever (Gin): A Spirit Full of History, Science, and Technology." *Sartonia* 12: 93–147.

Chapter 14. Rum (and Cachaça)

Broom, Dave. 2017. *Rum: The Manual.* London: Mitchell Beazley.
Curtis, Wayne. 2018. *And a Bottle of Rum, Revised and Updated: A History of the World in Ten Cocktails.* New York: Broadway Books.
Minnick, Fred. 2017. *Rum Curious: The Indispensable Guide to the World's Spirit.* Beverly, MA: Voyageur Press.
Moldenhauer, Giovanna. 2018. *The Spirit of Rum: History, Anecdotes, Trends, and Cocktails.* Milan: White Star Publishers.
Smith, F. H. 2005. *Caribbean Rum: A Social and Economic History.* Gainesville: University Press of Florida.

Chapter 15. Eaux-de-Vie

Apple, R. W., Jr. 1998. "Eau de Vie: Fruit's Essence Captured in a Bottle." *New York Times,* April 1. https://www.nytimes.com/1998/04/01/dining/eau-de-vie-fruit-s-essence-captured-in-a-bottle.html.
Asimov, Eric. 2007. "An Orchard in a Bottle, at 80 Proof." *New York Times,* August 15. https://www.nytimes.com/2007/08/15/dining/15pour.html.

"Hans Reisetbauer: Austrian Superstar of Craft Distilling." 2021. https://www.flaviar.com/blog/hans-reisetbauer-eau-de-vie-from-austria.

Chapter 16. Schnapps (and Korn)

Prial, Frank. 1985. "Schnapps, the Cordial Spirit." *New York Times,* October 27. https://www.nytimes.com/1985/10/27/magazine/schnapps-the-cordial-spirit.html.

Weisstuch, Lisa. 2019. "Following a Trail of Schnapps through Germany's Storied Black Forest." *Washington Post,* October 25. https://www.washingtonpost.com/lifestyle/travel/following-a-trail-of-schnapps-through-germanys-storied-black-forest/2019/10/24/1fb62076-f030-11e9-89eb-ec56cd414732_story.html.

Well, Lev. 2016. *800 Schnapps-based Cocktails.* Scotts Valley, CA: Create Space Publishing.

Chapter 17. Baijiu

Huang, H. T. 2000. *Biology and Biological Technology.* Part 5 of J. Needham, *Science and Civilization in China,* vol. 6. Cambridge, MA: Harvard University Press.

Kupfer, Peter. 2019. *Bernsteinglanz und Perlen des Schwarzen Drachen: Die Geschichte der chinesischen Weinkultur.* Deutsche Ostasienstudien 26. Großheirath, Germany: Ostasien Verlag.

McGovern, Patrick E., Fabien H. Toro, Gretchen R. Hall, et al. 2019. "Pre-Hispanic Distillation? A Biomolecular Archaeological Investigation." *Open Access Journal of Archaeology and Anthropology* 1 (2). doi: 10.33552/OAJAA.2019.01.000509.

Sandhaus, Derek. 2015. *The Essential Guide to Chinese Spirits.* Melbourne: Viking Australia.

Sandhaus, Derek. 2019. *Drunk in China: Baijiu and the World's Oldest Drinking Culture.* Lincoln, NE: Potomac Books.

Chapter 18. Grappa

Behrendt, Axel, Bibiana Behrendt, and Bode A. Schieren. 2000. *Grappa: A Guide to the Best.* New York: Abbeville Press.

Boudin, Ove. 2007. *Grappa: Italy Bottled.* Partille, Sweden: Pianoforte.

Lo Russo, Giuseppe. 2008. *Il piacere della grappa.* Florence: Giunti.

Musumarra, Domenico. 2005. *La grappa Veneta: Uomini, alambicchi e sapori dell'antica terra dei Dogi.* Perugia: Alieno.

Owens, Bill, Alan Dikty, and Andrew Faulkner. 2019. *The Art of Distilling, Revised and Expanded: An Enthusiast's Guide to the Artisan Distilling of Whiskey, Vodka, Gin, and Other Potent Potables.* Beverly, MA: Quarto.

Pillon, Cesare, and Giuseppe Vaccarini. 2017. *Il grande libro della grappa.* Milan: Hoepli.

CHAPTER 19. Orujo (and Pisco)

"All about Pisco." 2021. Museo del Pisco. http://museodelpisco.org.
Orujo from Galicia (official page, in Spanish). 2021. http://www.orujodegalicia.org.

CHAPTER 20. Moonshine

Hogeland, William. 2010. *The Whiskey Rebellion: George Washington, Alexander Hamilton, and the Frontier Rebels Who Challenged America's Newfound Sovereignty.* New York: Simon and Schuster.
Lippard, Cameron D., and Bruce E. Stewart. 2019. *Modern Moonshine: The Revival of White Whiskey in the Twenty-First Century.* Morgantown: West Virginia University Press.
Okrent, Daniel. 2010. *Last Call: The Rise and Fall of Prohibition.* New York: Simon & Schuster.
"Taxation of Alcoholic Beverages." 1941. CQ Researcher Archives. https://library.cqpress.com/cqresearcher/document.php?id=cqresrre1941022800.

CHAPTER 21. And for Good Measure . . .

Blue, Anthony Dias. 2004. *The Complete Book of Spirits; A Guide to Their History, Production, and Enjoyment.* New York: William Morrow.
McGovern, Patrick E. 2019. "Alcoholic Beverages as the Universal Medicine before Synthetics." Pp. 111–127 in M. V. Orna, G. Eggleston, and A. F. Bopp, eds., *Chemistry's Role in Food Production and Sustainability: Past and Present.* Washington, DC: American Chemical Society.
Miller, Norman. 2013. "Soju: The Most Popular Booze in the World." *Guardian,* December 2. http://www.theguardian.comlifeandstyle/wordofmouth/2023.dec/02/soju-popular-booze-world-south-korea.
Tapper, Josh. 2014. "Slivovitz: A Plum (Brandy) Choice." *Moment* (March–April). https://momentmag.com/slivovitz-plum-brandy-choice.

CHAPTER 22. Cocktails and Mixed Drinks

Archibald, Anna. 2021. "The Nine Most Important Bartenders in History." Liquor.com. https://www.liquor.com/slideshows/9-most-important-bartenders-in-history.
Aznar, M, M. Tsachaki, R. S. T. Linforth, V. Ferreira, and A. J. Taylor. 2004. "Headspace Analysis of Volatile Organic Compounds from Ethanolic Systems by Direct APCI-MS." *International Journal of Mass Spectrometry* 239 (1): 17–25.

Brown, Derek. 2018. *Spirits, Sugar, Water, Bitters: How the Cocktail Conquered the World.* New York: Rizzoli.

Buehler, Emily. 2015. "In the Spirits of Science." *American Scientist* 103 (4): 298.

Cipiciani, A., G. Onori, and G. Savelli. 1988. "Structural Properties of Water-Ethanol Mixtures: A Correlation with the Formation of Micellar Aggregates." *Chemical Physics Letters* 143 (5): 505–509. doi.org/10.1016/0009-2614(88)87404-9.

Déléris, I., A. Saint-Eve, Y. Guo, et al. 2011. "Impact of Swallowing on the Dynamics of Aroma Release and Perception during the Consumption of Alcoholic Beverages." *Chemical Senses* 36 (8): 701–713. doi.org/10.1093/chemse/bjr038.

Luneke, Aaron C., Tavis J. Glassman, Joseph A. Dake, Amy J. Thompson, Alexis A. Blavos, and Aaron J. Diehr. 2019. "College Students' Consumption of Alcohol Mixed with Energy Drinks." *Journal of Alcohol & Drug Education* 63 (2): 59–95.

Mencken, H. L. 1919. *The American Language.* New York: Alfred Knopf.

Niu, Yunwei, Pinpin Wang, Qing Xiao, Zuobing Xiao, Haifang Mao, and Jun Zhang. 2020. "Characterization of Odor-Active Volatiles and Odor Contribution Based on Binary Interaction Effects in Mango and Vodka Cocktail." *Molecules* 25 (5): 1083.

Qian, Michael C., Paul Hughes, and Keith Cadwallader. 2019. "Overview of Distilled Spirits." Pp. 125–144 in Brian Guthrie et al., eds., *Sex, Smoke, and Spirits: The Role of Chemistry.* Washington, DC: American Chemical Society.

Ranasinghe, Nimesha, Thi Ngoc Tram Nguyen, Yan Liangkun, Lien-Ya Lin, David Tolley, and Ellen Yi-Luen Do. 2017. "Vocktail: A Virtual Cocktail for Pairing Digital Taste, Smell, and Color Sensations." Pp. 1139–1147 in *Proceedings of the 25th ACM International Conference on Multimedia.* New York: Association for Computing Machinery.

Thomas, Jerry. 2016, reprint. *Jerry Thomas' Bartenders Guide: How to Mix All Kinds of Plain and Fancy Drinks.* New York: Courier Dover.

Wondrich, David. 2016. "Ancient Mystery Revealed! The Real History (Maybe) of How the Cocktail Got Its Name." *Saveur,* January 14. https://www.saveur.com/how-the-cocktail-got-its-name.

Chapter 23. The Future of Spirits

Rappeport, Alan. 2019. "Trump's Trade War Leaves American Whiskey on the Rocks." *New York Times,* February 12. https://www.nytimes.com/2019/02/12/us/politics/china-tariffs-american-spirits.html.

Guest Contributors

Miguel A. Acevedo, Department of Wildlife Ecology and Conservation, University of Florida, Gainesville, FL, 32611, USA
Sergio Almécija, American Museum of Natural History, New York, NY, 10024, USA
Angélica Cibrián-Jaramillo, CINVESTAV, El Copal, Irapuato-León, CP, 36824, Mexico
Tim Duckett, Heartwood Malt Whisky, North Hobart TAS 7000, Tasmania, Australia
Joshua D. Englehardt, Centro de Estudios Arqueológicos, El Colegio de Michoacán, CP, 59370, La Piedad, Mexico
Michele Fino, University of Gastronomic Sciences, 12042 Pollenzo, Italy
Michele Fontefrancesco, University of Gastronomic Sciences, 12042 Pollenzo, Italy
América Minerva Delgado Lemus, Manejo Integral y Local de Productos Agroforestales A.C., Mexico
Pascaline Lepeltier, Racines NY, New York, NY, 10007, USA
Christian McKiernan, New York, NY, USA
Mark Norell, American Museum of Natural History, New York, NY, 10024, USA
Susan Perkins, Division of Science, The City College of New York, New York, NY, 10031, USA
Bernd Schierwater, Institute of Animal Ecology and Cell Biology, University of Veterinary Medicine, Hanover, Germany
Ignacio Torres-García, Environmental Transdisciplinary Studies, Escuela Nacional de Estudios Superiores, Universidad Nacional Autónoma de México, Morelia, Mexico
Alex de Voogt, Drew University, Madison, NJ, 07940, USA
David Yeates, Canberra ACT, Australia

Index

Figures and tables are indicated by f and t following the page number.